Exploring Electronic Media

Exploring Electronic Media:

Chronicles and Challenges

Peter B. Orlik
Steven D. Anderson
Louis A. Day
W. Lawrence Patrick

Blackwell
Publishing

© 2007 by Peter B. Orlik, Steven D. Anderson, Louis A. Day, and W. Lawrence
Patrick

BLACKWELL PUBLISHING
350 Main Street, Malden, MA 02148-5020, USA
9600 Garsington Road, Oxford OX4 2DQ, UK
550 Swanston Street, Carlton, Victoria 3053, Australia

The right of Peter B. Orlik, Steven D. Anderson, Louis A. Day, and W. Lawrence
Patrick to be identified as the Authors of this Work has been asserted in accordance
with the UK Copyright, Designs, and Patents Act 1988.

First published 2007 by Blackwell Publishing Ltd

1 2007

Library of Congress Cataloging-in-Publication Data

Exploring electronic media : chronicles and challenges / Peter B. Orlik . . . [et al.].
    p. cm.
  Includes bibliographical references and index.
  ISBN-13: 978-1-4051-5054-5 (hardcover : alk. paper)
  ISBN-10: 1-4051-5054-8 (hardcover : alk. paper)
  ISBN-13: 978-1-4051-5055-2 (pbk. : alk. paper)
  ISBN-10: 1-4051-5055-6 (pbk. : alk. paper)   1. Mass media–Technological
innovations–History.   I. Orlik, Peter B.

  P96.T42E98   2007
  302.2309–dc22   2006027958

A catalogue record for this title is available from the British Library.

Set in 11 on 13 pt Bembo
by SNP Best-set Typesetter Ltd, Hong Kong
Printed and bound in Singapore
by COS Printers Pte Ltd

The publisher's policy is to use permanent paper from mills that operate a
sustainable forestry policy, and which has been manufactured from pulp processed
using acid-free and elementary chlorine-free practices. Furthermore, the publisher
ensures that the text paper and cover board used have met acceptable environmental
accreditation standards.

For further information on
Blackwell Publishing, visit our website:
www.blackwellpublishing.com

Contents

Illustrations

Figures

Tables

About the Authors

Steven D. Anderson is Professor in the School of Media Arts and Design at James Madison University where he conducts courses in electronic media technologies. His teaching and research specialties include streaming media, web development, and the history of media technology. He has received numerous national awards for his multimedia and video work and has published across a broad range of academic and trade journals. He is a past president of the Broadcast Education Association and served as its 2000 convention program chair. He also chaired BEA's Production Aesthetics and Criticism Division, co-chaired BEA's Creative Activities Task Force and served on the organization's Multimedia Task Force. Prior to entering academe, he was environmental reporter and weekend weathercaster for CBS's KCNC-TV (Denver) following previous stints as news photographer, weathercaster and news reporter at stations in Fresno, California and Fargo, North Dakota.

Louis A. Day is an alumni Professor in Louisiana State University's Manship School of Mass Communication. He has taught communications law and electronic media regulation for more than 30 years and has published legal research in such prestigious venues as *Communications Law and Policy*, *Communication and the Law*, *Journalism Quarterly*, and *Journal of Broadcasting and Electronic Media*. He is also the author of one of the most widely used media ethics texts: *Ethics in Media Communications: Cases and Controversies* (Wadsworth) and is a frequent presenter on media law and ethics at the annual conventions of the Broadcast Education Association and the Association for Education in Journalism and Mass Communication. His prior industry experience includes work as a reporter, writer, and editor for WRBL in Columbus, Georgia and the US Army in Vietnam.

Peter B. Orlik is Professor and Director of Central Michigan University's School of Broadcast and Cinematic Arts; a unit he founded in 1969. A former advertising agency copywriter, radio music programmer, and television creative services executive, he is the author of three current textbooks: *Broadcast Cable Copywriting* 7th edition (Allyn & Bacon); *Electronic Media Criticism: Applied Perspectives* 2nd edition (Lawrence Erlbaum Associates); and *Career Perspectives in Electronic Media* (Blackwell). He is a past member of the Broadcast Education Association's board of directors, two-time chair of its Courses, Curricula and Administration Division and has chaired BEA's Scholarship Committee since 1992. He was the 2001 recipient of BEA's Distinguished Education Service Award and, in 2003, became the first educator to be inducted into the Michigan Association of Broadcasters Hall of Fame.

W. Lawrence Patrick is President of Patrick Communications, LLC, a leading media investment banking and brokerage firm as well as owner of Legend Communications, a 16-station radio group. He previously served as senior vice president at the National Association of Broadcasters and later, COO of Gilmore Broadcasting. Patrick was a full-time professor at the University of Tulsa and the University of Maryland and now holds adjunct full professor rank at the Georgetown University Law Center and Central Michigan University. He currently serves as president of the National Association of Media Brokers, trustee of the Television and Radio Political Action Committee for the National Association of Broadcasters and member of the Michigan Association of Broadcasters Foundation board of directors. He is a former president of the Broadcast Education Association and recipient of the organization's 2005 Distinguished Education Service Award – BEA's highest honor.

Preface

We have entitled this book *Exploring Electronic Media* because, even though such media have been around in some form for well over a century, these vehicles and their output are changing and expanding as never before. Professionals who have worked in the field for decades still find each day to be a process of discovery – so people who are now seeking to gain entry to this profession should not feel over-whelmed at how much about it there is to learn.

The most successful explorations are purposeful as well as satisfying. Therefore, this book has been structured to provide the maximum amount of insight in a minimum of pages while offering revelations about the field that are both significant and useful. Learning your way around the media world is a process that, once begun, is never fin-ished. But we hope the reconnaissance conducted within the follow-ing pages will provide a suitable roadmap for further investigations.

Exploring Electronic Media: Chronicles and Challenges is divided into two Parts: Chronicles and Challenges. Part I starts in Chapter 1 by ori-enting you to the fundamental aspects of human communication and how these aspects impact the operation of mass and other media systems. The four Chronicles chapters subsequently investigate how these systems came to be as they are. Chapter 2 lays out technologi-cal development because it is technology that provides our delivery systems with their physical attributes and capabilities. Next, Chapter 3 examines how the media content distributed by this technology has evolved as it is this content that constitutes the core product of our enterprise and what is most obvious to consumers. With this understanding of media hardware and software, we can then proceed in Chapters 4 and 5 to analyze how regulatory and business forces have shaped and determined media conventions and functions.

These historical discussions are designed to help you comprehend how the electronic media landscape has evolved to its present state. Knowing that background, it is easier to fathom the forces now shaping our media environment and to more accurately appraise where those forces might lead. Pat II, Challenges (Chapters 6–9) performs this appraisal by retracing the progression utilized in the Chronicles segments. But now, the focus is on the *current* issues, controversies, problems, and potentials faced in media technological content, legal and economic arenas respectively.

Each of the nine chapters concludes with a summarizing *Rewind* section, followed by *Self-Interrogation* questions designed to probe understanding and confirm what your reading explorations have discovered. Having mastered this material, it is our hope that you will be well equipped to continue your electronic media journey through subsequent readings and experiences in the years ahead. Thank you for permitting us to begin this expedition with you.

Pete Orlik
Steve Anderson
Lou Day
Larry Patrick

Acknowledgments

This book is not the work of just four authors, but also the product of many family members and colleagues who provided their support. Thanks to Chris for her love and constant nourishing of this and many other book projects and to Jay Rouman for rescuing the manuscript by building it a new computer. Thanks also to Mary Ellen for her love and patience, to computer engineers Matthew and Erik for their never-ending curiosity and for keeping their father up on technology, and to Kyle for helping provide the laughter. Appreciation is also expressed to Michele Cooney and Summer Foust who ably assisted in the preparation of Chapters 5 and 9. Finally, the authors thank Elizabeth Swayze for championing the conception, and execution, of this exploration of the electronic media profession to which all four of us have devoted our professional lives.

CHAPTER 1

Bringing Electronic Media into Focus

Peter B. Orlik

Before we delve into the electronic media's specific chronicles and challenges, it is important to understand how these media fit into the overall communication matrix. For despite the prominent and sometimes breathtaking power of electronic technologies, they still must be harnassed to serve age-old, people-centric tasks.

Components of Communication

We have all heard the term *communication*. Likewise, we have all been exposed to the concept of *mass communication*. But exactly what do we mean by these terms? Where does one term leave off and the other begin? Are they mutually inclusive or mutually exclusive? And how do they relate to the work and character of the electronic media? We might say, for example, that the electronic mass media are vehicles for achieving mass communication, but this statement gets us nowhere if we do not understand what such achievement entails.

As the first step in making sense of all of this, let us begin with the following basic presumption:

Communication among human beings is the process of conveying and exchanging information, assertions, and attitudes.

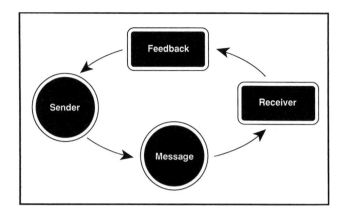

Figure 1.1   The communication loop

Because this is not a book on horticulture or animal husbandry, we are not interested in how plants or animals converse. Nor are we computer scientists plotting interactions between banks of silicon chips or psychiatrists seeking to explain how people relate to their inner selves. Instead, we are simply (or not so simply) concerned with what happens in the establishment of linkages *between people*, or, as the definition puts it, "among human beings."

Also, we must realize that such communication is an ongoing *process*; it is not the flipping of a switch or a single burst of data, but an ongoing stream of interactions to which both the initiator and receptor of the initial cue contribute. Every successful communicator discovers that receivers are more than passive sponges who merely soak up data. Rather, these receivers are communicators in their own right who might respond in a variety of ways to the apprehended message. In fact, the process of human communication is so circular, so ongoing, that it is sometimes difficult to detect which party initially set the activity into motion. Even though we tend to refer to the initiating stimulus as a *message*, and to the receiver's response as *feedback* (see Figure 1.1), it is often difficult to distinguish one from the other.

For example, little Billy tells his classmate Abby to "buzz off" or "get out of my face." Is this the initial message or is it feedback in response to a perceived gesture or verbal comment from her? A third party (such as the teacher) might not be able to figure this out and, perhaps, neither will the two children. Abby's earlier gesture or comment may not even have been directed to Billy but to another member of the class. Consequently, what Billy intended as feedback is misinterpreted by Abby as an unfriendly introductory message. Communication is a *process*, a loop, the beginning and end of which may defy easy identification.

As this example suggests, the *transmitting* function of communication can be both verbal and nonverbal. We communicate with posture, gestures, touches, and facial expressions as well as with sounds and words in order to bridge the physical and/or psychological distance between people. Sometimes, as in the case of a friendly wave across a parking lot or the typical telephone conversation, we rely exclusively on either nonverbal or verbal communication. More frequently, our linkage with another person is a blending of these two modes. Usually, such a combination strengthens the communication bond. But when the verbal seems to contradict the nonverbal – when the words say "It's true" while the eyes shift away – the transmission becomes more complicated to decode, though no less meaningful.

Whether verbal, nonverbal, or both, our messages transmit three categories of content, and they are often conveyed simultaneously. As our definition indicates, this content trio consists of: (1) information; (2) assertions; and (3) attitudes.

Taken by itself, *information* is simple conveyance of fact. "The flight will leave at 9:35"; "Santa Claus was a bishop called St. Nicholas who died in the fourth century"; "Lima beans make me puke." These are three seemingly straightforward pieces of knowledge that one person might share with another. Yet each item, under certain circumstances, may also be perceived as a debatable *assertion*. Coming from a harried gate agent, the promise that the flight will leave at 9:35 may be taken more as hope than certainty, particularly by passengers who have already been twice delayed. Santa/St. Nick probably is departed, but that is of doubtful credibility to the Christmas Eve child who thinks she hears reindeer droppings hitting the roof. And despite your distaste for lima beans, your physician may disagree with your contention that they possess the chemical capacity to make you nauseous. So, depending on the receiver of the communication, each of these messages may be perceived, not as objective information, but as subjective assertion.

Finally, *attitude* also comes into play in any human communication transaction because it is impossible for people either to send or receive a message without encasing it in an opinion-suggesting wrapping. People are not robots. They make evaluative and emotional judgments on every subject they encounter and look for signs of such judgments in the transmissions of others. If such a sign is difficult to detect, they might conclude that the other person is either devious or insensitive. Thus, the gate agent's scrupulously flat recitation of the flight's alleged departure time may be seen as a deliberate cover-up by some people and as cold indifference by others. The recited fact (or assertion) of Santa's demise might strongly imply that the reciter is callous and oblivious to childhood's joys. And the slovenly slang chosen to describe the lima beans' effect will convey as much about the speaker's appraisal of the listener's sensitivities as it does about the vegetable's impact. For practice in communication analysis, examine the commercial in Figure 1.2. What elements of this message are information? Assertion? Attitude?

The remoteness of *written* words on a page may make an assertion seem like unadorned fact and disguise the attitude behind it. *Spoken* communication, however (including that carried via television and radio), conveys many more judgmental clues. Vocal tone and pitch, length of pauses and speed of delivery, eye contact, posture, facial expression, and gestures all convey intended or accidental hints as to the feelings and beliefs that accompany and motivate the statement. These hints may be distortive or inaccurate, but they will be taken into account nonetheless. Inevitably, interpersonal communication is *expected* to convey information, assertions, and attitudes. If a given message seems devoid of one of these factors, we as receivers are likely to supply that factor from our own storehouse of presumptions gathered from past human interactions.

Another way to define the communication task is to assert that "communication is the carrier of the social process." This is certainly a task statement, but it doesn't tell us much unless we know what that conveyed social process involves. To pioneering communication theorist Harold Lasswell, the phenomenon consists of three major components: "(1) the surveillance of the environment; (2) the correlation of the parts of society in responding to the environment; (3) the transmission of the social heritage from one generation to the next."[1] The first two elements can be accomplished by any living entity. The third

Figure 1.2 A piece of television-assisted human communication. Source: Courtesy of FCB/Leber Katz partners.

separates human beings from lower life forms and therefore constitutes the crucial and unique ingredient of the human communication task.

In a plant, the taproots survey for moisture. If the eastern-most root detects it, the correlated response directs more root capacity in that direction. A member of a flock of sparrows looks around and sees a cat ready to pounce on them. The flock's correlated response is to fly to a safer locale. Yet future generations of plants and sparrows do not move beyond these simple reactions. Plants do not learn to engineer irrigation systems to bring themselves distant water, and flocks of sparrows do not construct cat traps. Each succeeding generation begins at about the same place as previous ones, with only the lengthy process of biological evolution to offer any chance of enhanced adaptation.

Through their possession of advanced communication skills, however, human beings can achieve that vital third step of *transmitting social heritage*; they record and transmit new methodologies and insights so that each generation can build on the advances made by its immediate predecessor. Unlike the lower life forms, human beings have a history that, when conveyed by human communication, has the capacity to promote often rapid generational progress and development. Thus, this component of social process not only separates human communication from that practiced by lower life forms but it also defines and enhances humanity itself.

This is why the National Council of Churches of Christ, in a major policy statement entitled *Global Communication for Justice and Peace*, asserts that "The right to communication is a basic human right. The right to receive and to provide information is as fundamental to the quality of life as food, clothing and shelter. The right to communicate is essential to human dignity."[2]

In its simplest form, communication's transmission of social heritage is achieved through the memorization of truths, principles, and skills that are then orally passed along from person to person. Often, these insights are put in the form of songs, poems, and stories (fables) to make memorization and recitation easier. The biblical psalms, for example, were perfect for an overwhelmingly nonliterate society. They required neither the ability to read nor the then-considerable expense of writing equipment in order to be conveyed. Instead, their lyrical imagery made the task of memorization a pleasant and attainable accomplishment. Often, as in the case of the *rhapsodes* of ancient Greece, a whole profession was created to facilitate such heritage transmission via poetry. "As this poetry becomes more popular and addresses itself to a wider public," art historian Arnold Hauser points out, "so the style of delivery becomes less and less formal and approaches that of everyday speech."[3] Thus, with the psalmists, rhapsodes, and similar ancients from all the world's cultures, we may confront the conversational ancestors of today's electronic media professionals.

And just as the modern electronic communicators' world is competitive, so too was that of their Greek predecessors. Classics scholar Mark Edwards reveals that many poems dating from around the time of Homer evidence a high level of competition among poets. Some of these poems talk of contests among the singers/poets for audience attention – not unlike the ratings races of today. In other poems, their creators proclaim how they have corrected information in the stories conveyed by their rivals – akin to today's competitive posturings of television news departments and tabloid producers.[4]

But as the more perceptive ancient communicators probably recognized, passing social heritage solely through face-to-face encounters is an inefficient and time-consuming enterprise. The number of psalmists, rhapsodes, minstrels, and similar history passers available to any society is exceedingly small compared with the population to be served. To cope with this shortage, more advanced societies further enhance themselves through the development and use of *communications* (with an *s*).

Communications Vehicles

In a nutshell, communications are implements that enable us to extend our ability to interact over time and space. Unlike the rhapsode's voice, these vehicles are not part of the human anatomy. Instead, they are separate tools that allow us to communicate in situations in which face-to-face contact is not possible.

At 8:00 a.m., a teacher writes an assignment on the chalkboard for a 10:00 a.m. class and then leaves town on an emergency errand. Even though that absent instructor may be more than a hundred miles away by the time the 10:00 a.m. students arrive, she still has communicated with them. Through use of as simple a "tool set" as a piece of chalk and a sheet of slate, the instructor has managed to leap both time and space in completing her transmission. In a modern society, such communications tools are all around us and are readily available. The telephone-answering device and computer auto-reply function represent the more sophisticated of such mechanisms, and the note on the refrigerator or pile of rocks that mark a forest trail constitute more primitive communications vehicles. Yet, because all of these methods extend our ability to disseminate over time and distance, all are powerful out-of-body amplifiers of the communication process. They are technical implements through which the impact of our information, assertions, and attitudes can be magnified.

Communications vehicles therefore possess the associated capacity to build *dispersed communities*. A second-century Roman physician may have resided hundreds of miles from other healers, but through common access to written scrolls (technical implements) detailing certain medical procedures, he could join other such healers to create a community of physicians, even though most would never meet each other face-to-face. Despite today's vastly improved transportation systems, trade and professional communities of all types still owe their existence, not to eyeball-to-eyeball interchange, but to the more or less continuous technological linkages that communications vehicles provide.

Obviously, a great variety of such linkage tools are currently available. As we begin this book's exploration of the electronic media, however, we need to focus our primary attention on the category of *professionally operated* communications tools that offer the greatest potential for communication carriage. This category is known as *mass communications* (again with an *s*), what Professor Edwin Emery dubbed the "intricate, many-faceted machinery"[5] that provides the technological wherewithal to proficiently deliver information, assertions, and attitudes to vast segments of the population. More precisely, *mass communications* can be defined as "those hyper-efficient and highest capacity communications tools that permit timely transmittal to the largest number of dispersed individuals."

Clearly, we are talking now about more than the chalkboard or the telephone. True mass communications mechanisms give us the likelihood of reaching millions of people at roughly the same time. And, even though captured audience members may share similarities of age, occupation, or locale, they are just as likely to be highly diverse in these and other characteristics. In short, it is the fundamental *mass*iveness of mass communications tools and their achievable audiences that sets these delivery systems apart from other, less potent communications vehicles.

Up to this point we have distinguished between *communication* (a process) and communication*s* (the tools that can extend this process over time and space). We have also isolated *mass* communications as the *professional* tools possessing the greatest efficiency and capacity. What remains is to delineate *mass communication* (no *s*), the process made possible only through the use of mass communications devices.

For comparison and consistency, therefore, let us modify our original definition of *communication* and propose that *mass* communication is

> the process of rapidly conveying *identical* information, assertions, and attitudes to a *potentially large*, *dispersed*, and *diversified* audience via mechanisms capable of achieving that task.

There are obviously some key differences between this statement and our earlier communication definition. Five of these differences relate to the addition of the relative terms *identical, potentially, large, dispersed*, and *diversified*. Let us look at each requisite separately.

When something is *identical*, of course, it is exactly the same as another entity with which it is being compared. Real mass communication offers this "same as" quality. Even though 50,000 different people may hear a radio station's early morning newscast, they are all hearing exactly the same material arranged in exactly the same way. Similarly, even though 20 million different individuals may watch a particular network television drama this evening, and watch it over multiple time zones, they are all being exposed to precisely the same program content as it has been preserved on film, tape, or digital stock.

On the other hand, if you separately meet three friends on the street and tell each the new joke you've just heard, these three retellings will not be identical. No matter how hard you try to be consistent, you will use different words, phrases, and gestures each time you relate your witticism. There will be even less uniformity, of course, if you tell the story only to the first friend, who passes it on to the second, who, in turn, conveys it to the third. In this case, different communicators as well as different phraseology will impact the joke's transmittal, with the subsequent likelihood of some very unidentical messages, including at least one that may mangle the punchline.

Through its ability to make a virtually unlimited number of impressions of the exact same original, however, mass communication can transmit identical content to every person in a position to receive the message. Whether these people: (1) *decipher* the message identically; and (2) *respond* identically is another matter, of course. As we've mentioned, each person is a unique receiver whose feedback (or lack of same) constitutes a completely individualized, discrete, and often unpredictable response.

Our second relative term, *large*, is more ambiguous. We can come to grips with it only by taking a "compared with what?" orientation. Two thousand people in a theater watching a new musical comedy is a large crowd compared with 200, but not compared with the 200,000 who might easily be exposed to the play if it were telecast over a noncommercial broadcasting network. A politician's address to a trade union convention may reach 1,500 delegates, but how much larger will the audience be if the address is reprised on a cable news network? Clearly, there is a quantum leap in what constitutes large once we recruit one of the mass communications vehicles to transmit our consequently *mass* message.

This "large" ambiguity is especially apparent when we consider the Internet. A given website certainly has the global potential to reach even more people than a domestic television network's prime time broadcasts. However, with the millions of websites from which consumers may choose, it is only those mounted by large *professional* online enterprises (like Yahoo!) or affiliated with more traditional electronic media and entertainment companies (like sites for networks, stations, music labels and pro sports leagues) that have the marketing and cross-promotional wherewithal to aggregate mass-communication-size audiences. As early as 2002, Professors James Webster and Shu-Fang Lin found that "the top 200 sites alone account for roughly half the traffic on the Web."[6] These sites may attract mass-size audiences – but usually as partnerships between "new" media and "old".

Our third relative term, *dispersed*, pertains to the location of the audience members. A cavernous auditorium may be packed with a comparatively large crowd made up of many diverse types of people who are all receiving an identical message from the speaker or performance on stage. Yet, because these people are all in the same locale, they are not dispersed. All were required to travel to the event in order to experience it. The process of mass communication, in contrast, brings the event to the audience rather than the other way around. This fact provides great flexibility in both group size and group location.

We could, for example, assemble substantial numbers of viewers to watch educational or political indoctrination programs on community television sets in their respective neighborhoods, something regularly done in many Third World countries. At the other extreme, we can follow the standard US practice of reaching millions of geographically separated families or even solitary viewers on their own personal receivers. Either way, our audience is dispersed, and the size of each viewer group is less important than the fact that the televised event was actually delivered simultaneously to all of these groups.

The *diversified* element in our mass communication definition provides our fourth comparative, though imprecise, benchmark. Even though the 20 million people from our earlier example who watched that network drama certainly represent a broader spectrum of society than do the 200,000 who viewed the musical comedy on public television, both audiences are much more variegated than is the populace attending that trade union speech or the three friends exchanging jokes. To a considerable degree, the larger a group, the more diversified it is likely to be. Nevertheless, the factors of audience size and composition must be considered separately in gauging whether a given situation possesses the potential for mass communication.

Finally, the word *potential* (or *potentially*, as it appears in our definition) is extremely important. There are times when a mass communicator deliberately chooses *not* to use the full mass communication capacity that a vehicle offers. *Broadcasting & Cable*, for example, is a US trade magazine that seeks to serve professionals working in, or offering services for, the commercial radio and television industries. It is disseminated almost entirely by subscription to members of this target audience with an average weekly circulation of 24,000 copies. Compared with other national magazines such as *Time* or *People*, *Broadcasting & Cable*'s readership is not particularly large and is certainly not diversified. Instead, the publishers of *Broadcasting & Cable* (and of many other specialty magazines) have made the conscious decision not to exploit the national magazine format's *potential* for a sizeable and heterogeneous audience. They closely control circulation so that advertisers seeking exclusively to reach people in the radio/television business will be attracted to *Broadcasting & Cable*. These advertisers know that they are paying to contact only individuals who might actually be in a position to use their products or services with no "waste circulation" among uninterested readers.

Using the same printing press, manufacturing, and basic mail distribution process, *Broadcasting & Cable* could decide to *function* as a mass communications vehicle simply through content changes. The articles on leveraged buyouts of cable systems could be replaced with discussions of sexual perversion in suburbia, and the two-page comparison of video servers could be exchanged for a thong-clad centerfold with a staple in the navel. What would have changed? Not the *form* of the magazine, but rather its *content*. It is important to recognize, therefore, that while content determines whether a given entity is actually functioning as a system for mass communication, it is *form* that decrees whether such potential could exist. In other words, *Broadcasting & Cable* (and, in fact, any trade publication or niche cable network) is a mass medium in form that chooses not to fully exploit that mass communication ability in its featured content.

The Broadcast and Nonbroadcast Electronic Media

Having defined communication and mass communication activities, we can orient ourselves more precisely to our primary area of interest – *the electronic media*. As an industry, these media generate and disseminate a signal – over the air and/or down a wire – to permit and actively encourage timely transmittal to a large number of dispersed individuals. This description currently isolates radio and television enterprises as well as those Web-based companies that have deliberately positioned themselves to aggregate audiences for commercial, public relations, political or social purposes.

In the most traditional electronic mass media application (see Figure 1.3), a radio or television station catapults its message into the air over an assigned frequency. This broadcast communication is then capable of capture by all receiving sets within range of the signal for the benefit of the listening/viewing public.

The term broadcast is an appropriate descriptor of this process even though that word did not originate in the mass media field. To broadcast originally meant "to strew seeds in all directions," as a farmer would do in planting a freshly tilled field. Today's broadcast professionals are engaged in planting and germinating ideas, assertions, and attitudes rather than corn or wheat. Yet the concept of scattering the commodity in all directions remains as relevant to these electronic endeavors as it does to agriculture. In neither case does all that you sow bear fruit.

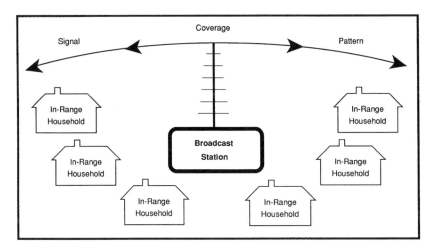

Figure 1.3   Basic broadcast signal transmission

But if you harvest too few plants (or too few audience members), you will not realize the profit necessary to continue your work. Of course, your efforts could be partially subsidized by the government – as in crop price supports or public broadcasting. Such schemes do not alter the fundamental nature of your business, however; they only modify the sources and measures by which your labors are compensated.

As our "over-the-air" and/or "down-a-wire" description implies, electronic communicators often use *nonbroadcast* delivery systems or combine such systems with broadcast applications. A cable network like CNN or ESPN, for example, is entirely nonbroadcast. Instead of scattering their signals in all directions over the open airwaves, cable networks use super high frequencies to uplink their signals from their main control rooms to communications satellites that then downlink them to the receiving dishes owned by individual cable systems. The signals are then downconverted to normal broadcast frequencies and distributed closed-circuit by a coaxial or fiber optic line only to viewers who have contracted and paid for the service.

Roughly the same thing happens in a DBS (direct broadcast satellite) situation. In that instance, however, the cable system is eliminated because the satellite signal is received and downconverted at each subscriber's residence rather than channeled and refined by an intermediary cable company.

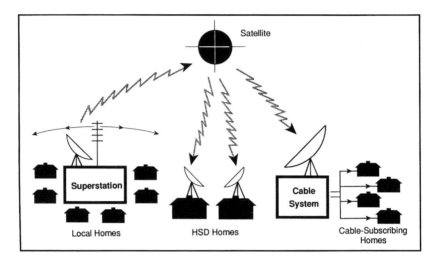

Figure 1.4   A superstation's combined broadcast/non-broadcast signal transmittal

A combination of both over-the-air (broadcast) and down-the-wire (nonbroadcast) technologies is represented by superstations such as Tribune Broadcasting's WGN in Chicago. In these operations, the station provides normal broadcast coverage to homes in the city to which it is licensed. At the same time, its signal is also being uplinked to a satellite for distribution to distant cable systems that offer it as another programming service to subscribers (see Figure 1.4).

The same signal may also be marketed directly to home satellite dish owners who are located outside a cable system's service area. These HSD (home satellite dish) homes pay a fee for the signal to an authorized dealer who also provides the descrambler necessary for the receiving set to make sense of the picture. In such a case, HSD homes are referred to as TVRO (television receive-only) installations to distinguish them from DBS. A basic TVRO dish merely intercepts signals being generated by and for other program delivery systems (such as broadcast and cable networks). A significantly smaller DBS dish, in contrast, picks up packages expressly intended for subscribers to the DBS enterprise as well as broadcast and cable content that the DBS operator has secured approval to carry. As the DBS industry has expanded, the number of TVRO installations has shrunk. Today, TVRO is usually found in locales unprofitable for DBS dealers to service or used by businesses to relay proprietary data and corporate programming.

Most recently, portable communications devices such as cell phones have acquired wireless broadband capabilities that enable them to carry audio and video programming to what is labeled "the third screen." In these cases, a variety of wireless providers serve as the intermediaries between program suppliers and portable-device-carrying consumers. Such devices are also further lessening the traditional electronic mass media limitation of facilitating only one-way communication.

Obtaining Feedback

Even though electronic communicators have long possessed the capacity to reach large, diverse, and dispersed audiences in an especially timely (even instantaneous) manner, our message transmission systems did not until recently permit receipt of immediate feedback from these audiences. Instead, radio and television professionals have had to wait for such delayed and limited responses as ratings reports, in-store customer traffic counts following an advertising blitz, and product sales computations. Programs and advertising campaigns are modified or replaced as a result of such subsequent feedback, of course. But unlike face-to-face communication, there has been little to help us monitor our success during or near the moment of actual message delivery. If we have miscommunicated, this fact has not usually been ascertained until days or even weeks after the occurrence.

Over the years, some limited exceptions to this one-way condition evolved. The radio talk show, for example, seems to offer a rich vehicle for two-way mass communication – until we examine its real purpose. Certainly the host of the program is conversing directly with the listener/caller. But that host's primary job is to make what the caller says of interest to the audience in general. Host and caller may be having an extended and seemingly productive interchange. Yet, if that conversation's subject or direction bores the rest of the listeners, the program will not hold its audience. At the time, the talk show host has no feedback from this audience *as a whole*. Thus, the host can rely only on professional experience and intuition as to whether this call is helping the program succeed.

Similarly, many television situation comedies, talk shows, and game programs are taped or filmed "before a live studio audience."

Theoretically, the feedback that performers receive from this audience should be a guide as to how viewers at home are reacting. In reality, however, that studio audience has been subjected to much additional stimulation not present in the home. Applause signs, an exuberant preshow warm-up by program assistants, and the very excitement of being in a real television setting can generate much more enthusiastic feedback from the studio audience than is being felt by home viewers on their living room couches. The producer who relies solely on the captive studio audience's reaction may be unpleasantly surprised when, weeks later, a ratings weakness or decline evidences a much different response on the part of the general viewing public.

But now lower equipment costs and expanding digital technology have made possible a wider range of interactive (two-way) experiences. As early as 1995, for example, Nova Television, the Czech Republic's first private TV station, set up a number of miniature recording booths around the city of Prague. By depositing the equivalent of 30 cents, customers could step into one of these little gray boxes and automatically record a short videotaped statement about what they felt to be some pressing issue. These statements were then featured by Nova Television on *Vox Populi*, a one-minute program airing several times throughout the station's program day. *Vox Populi* soon became one of Nova's most well-known offerings.[7]

At about the same time, and even more interactively, US program providers such as Warner were building websites in order to fashion two-way ties with their viewers. One Warner web area was set up to provide viewers with greater involvement in the studio's syndicated magazine show, *Extra*. Via this area, people could download photos of favorite guests, preview movie openings, submit questions to star performers, and take part in special polls, the results of which were aired on the program. Warner staffers then digested the thousands of e-mail responses received each week into 15-page packets that were distributed to *Extra*'s producers, providing what Warner executive and pioneering web marketer Jim Moloshok called

[a] grass roots report card on the program . . . As shows get bigger and more sophisticated, they tend to become less connected to the people viewing them at home. You are inclined when you are located in a city like Los Angeles to become very incestuous and convince each other that what you're doing is right. And anytime you can create a

connection to the end user of your product . . . it gives you instanta-neous feedback that what you're doing is right or wrong.[8]

Due to the ever-widening penetration of the Internet, a huge number of such interactive ventures have emerged from these and similar pio-neering efforts, providing additional exceptions to the historically one-way nature of mass communications. Viewers can now electronically participate in their favorite game show, vote in polls conducted as part of newscasts, and even choose the camera angle from which to view a developing football game. Through such things as VOD (video on demand) technology, they are empowered to select programming when they want it and thereby opt in to receiving commercials especially geared to their lifestyle preferences. Nevertheless, many electronic media consumers are not yet ready to forego the pleasures of passive viewing for the work of active keyboard or keypad participation. This ensures that at least the *mass* enterprises of the electronic media are likely to remain primarily one-way activities for some time to come.

Local/Network Dynamics

Initially, of course, electronic communication was all two-way via the telegraph's exchange of coded messages and the telephone's exchange of voiced ones. Near the turn of the twentieth century, the wireless apparatus then eliminated the necessity of stringing lines to connect the two communicating parties. However, once radio pioneers discov-ered there was money to be made in exclusively one-way broadcasts to multiple receivers, the electronic *mass* media were born. In deliver-ing content to consumers, these 1920s electronic communicators learned that aggregating both programming and delivery outlets was the best way to enlarge audiences and profits. The *network* concept came to the fore as the most efficient way to service a one-way com-munication relationship using local outlets as a vehicle for regional or national distribution.

The local/network pairing is somewhat akin to the manufac-turer/retailer relationship that provides consumers with everything from automobiles to soda pop. A broadcast or cable network *manufac-tures* (or contracts others to manufacture) the programming product.

This product is then distributed to the local retailers (the stations and systems) who make this product available to individual program consumers. Like many store franchises, this local outlet sometimes is owned by the company itself. In broadcasting, for example, stations owned by the network they carry are known as *O & Os* (owned and operated facilities).

From the beginning, broadcast media in the United States reflected local rather than national orientations, even as networking became well entrenched. While European countries, for example, were building highly centralized state systems exclusively to distribute national programming, the United States was allowing a host of individual stations and operators to serve their local communities in their own ways. Many of these stations later voluntarily joined to share programs (to *network*), but individual responsibility and decision making remained at the local level. When a US broadcaster receives a license from the Federal Communications Commission, that license is tied to a local community that the broadcaster is expected to serve. Even when carrying programs from other sources (such as networks), the local station is still held responsible for content decisions. Although it signs a contract with the network to distribute network programming, that local station (even if network-owned) can never, in theory, be required to air a network show that on-site station management believes to be contrary to the interests of its local community. US cable systems exhibit a similar local orientation. Usually, the local government (municipality, village, township) grants the cable company a franchise to operate and must decide whether to renew or terminate that franchise. In theory, the basic responsibility of the cable system thereby remains anchored to its community rather than to any of the national programmers with whom it does business.

Despite the fact that individual stations have been gathered together as station *groups*, and cable systems have largely consolidated into a relatively few giant MSOs (multi-system operators), such collective ownership does not change their legal responsibility to serve the local communities in which these outlets operate. However, some would argue that deregulatory trends of the past two decades in the United States and elsewhere have seriously eroded localism and made it increasingly difficult for communities to influence the operations of the huge media companies now likely to own the electronic outlets that serve their locality.

DBS would seem to constitute our only electronic mass media system that has never been grounded in the concept of localism (although the "less-mass" Web may be taking the same approach). This is true because, with DBS, the only local presence is the small satellite receive-only dish in the possession of each property owner. Programming is distributed on a national or even international basis and thus cannot easily be customized to provide local content. Therefore, it is unlikely that a DBS service itself can supplant the specific community orientation of a local broadcaster or cable access channel programmer, even if some high-capacity DBS satellites can now retransmit local station signals. For stationless regions of developing countries or for countries that have known only a national/centralized service, however, DBS offers significant cost advantages by eliminating the need to build an expensive string of terrestrial television facilities from scratch. On the radio side, satellite-delivered DARS (digital audio radio satellite) services promise similar cost advantages but localism drawbacks.

Network companies may be broadcast or nonbroadcast, free or pay, commercial or noncommercial, and private or publicly held, but the one thing they all have in common is their ability to share programming (and hence, program *costs*) efficiently with a number of individual outlets. The most prominent networks are national in scope, but international networks have long been the rule in shortwave radio and are coming to be the standard in European, Asian and most recently, Latin American DBS. At the other end of the network spectrum, state and regional networks make possible program exchange tailored more precisely to the needs and interests of a smaller geographic area. College and professional sports teams frequently set up networks in their home regions to carry their games, and state news nets are cost-effective ways of sharing regional information programming. A number of urban cable systems, meanwhile, have joined together in regional *interconnects* – not to share programming but to allow advertisers to make a single time buy that includes several adjoining cable operations.

The ability to facilitate national advertising through a single transaction has always been a fundamental service of, and reason for, commercial broadcast networks. Some media executives, in fact, would argue that the common carriage of the commercials is a more central function than is the delivery of the programs these commercials make possible. *Unwired networks* carry this idea to its ultimate conclusion. They put together large lists of stations on which they sell time. The unwireds do not deal in programming, however. Advertiser clients of unwireds thus accept the fact that their commercials will air on all of these stations at somewhat different times and adjacent to many different programs but often at significant cost savings over what the conventional full-service networks would charge.

Cable networks meanwhile may obtain their principal revenues from: (1) separate viewer payment for a particular "pay" service; (2) a cents-per-subscriber-per-month levy on cable systems that carry the network on their basic tier; (3) conventional national advertising transmitted with their programming; or (4) some combination of these three revenue streams. Like their broadcast counterparts, cable networks are designed to provide efficiencies of scale for everyone involved – programmers, advertisers, local media outlets, and consumers, who are the ultimate beneficiaries of the information and entertainment services that such efficiencies make possible.

Chapter Rewind

Though they are technologically sophisticated, the successful operation of electronic media systems still must be grounded in fundamental principles of human communication. This communication is the ongoing process of conveying and exchanging information, assertions, and attitudes. The process usually includes both verbal and nonverbal components which may inadvertently contradict each other. Human communication is also the carrier of *social process*, which involves the constant surveying of our environment, formulating responses to perceived environmental conditions, and then passing on successful responses to other people as heritage.

To extend their ability to communicate over time and space, societies develop communications vehicles. Some of the most basic of these tools are the chalk and slate board, the note on the refrigerator, or the pile of rocks that mark a wilderness trail. *Mass* communications vehicles are the hyper-efficient and highest-capacity communications mechanisms that permit timely transmittal to the largest number of dispersed individuals. Such vehicles – most notably the electronic media – are used to accomplish *mass communication* which entails rapidly conveying identical information, assertions, and attitudes to a potentially large, dispersed, and diversified audience. The electronic mass media can be divided into broadcast and nonbroadcast mechanisms. Broadcast transmissions are generally accessible as over-the-air signals by all receivers within each station's coverage patterns; nonbroadcast systems deliver their content only to paid subscribers via wire and/or over-the-air scrambled or password-protected messages.

Beginning as two-way, point-to-point endeavors, electronic media expanded into one-way *broadcast* communication endeavors to efficiently aggregate mass audiences. Recently, Internet and other digital technologies have offered the promise of increased interactive communication between electronic media companies and the consumers they seek to reach. One-way broadcast and cable enterprises have prospered within a local/network structure. Local outlets service a single community, whereas networks provide the same programming (and advertising efficiency) to a large number of outlets throughout a region or country as a whole. Radio and television in the United States have a long tradition of media localism that has somewhat moderated network power. With vehicles like DBS and the Web, however, it is now possible for a network to do without local outlets to reach people across national and even continental boundaries.

SELF-INTERROGATION

1   Compare and contrast *communication* and *communications*.
2   Compare and contrast *communications* and *mass communications*.
3   How does social process separate human beings from lower life forms?
4   What are the key characteristics of *dispersed* and *diversified* audiences?
5   Is it possible to achieve mass communication without using mass communications? Explain why or why not.
6   What is the key distinction between a *broadcast* and *nonbroadcast* activity?
7   Why were networks developed?

NOTES

1   Harold Lasswell, "The Structure and Function of Communication in Society," in Lyman Bryson (ed.), *The Communication of Ideas* (New York: Harper and Brothers, 1948), 38.
2   "Communicating for Justice: Document Articulates Right to Communicate," *Media and Values* (Winter 1993), 16.

3   Arnold Hauser, *The Social History of Art*, vol. 1 (New York: Vintage Books, 1951), 63.

4   Ellen Coughlin, "Research Notes," *Chronicle of Higher Education* (March 6, 1991), A6.

5   Edwin Emery, Phillip Ault, and Warren Agee, *Introduction to Mass Communications*, 4th edn (New York: Dodd, Mead and Company, 1974), 4.

6   James Webster and Shu-Fang Lin, "The Internet Audience: Web Uses and Mass Behavior," *Journal of Broadcasting & Electronic Media* (March 2002), 9.

7   Kristin Hohenadel, "Nova's Czech TV Show Stars the Populace," *Advertising Age* (January 16, 1995), I–22.

8   John Eggerton, "Jim Moloshok: Warner's E-Mailman," *Broadcasting and Cable* (February 13, 1995), 30.

PART I

Chronicles

CHAPTER 2

Technological Chronicles

Steven D. Anderson

As we saw in the previous chapter, the focus of this book is on communication between human beings to convey and exchange information, assertions and attitudes. While people have been communicating for thousands of years, it took many important technological innovations to create the wide capacity and widespread communication avenues we use today.

Early Communications Breakthroughs

Communicating without wires pre-dates wired communication by thousands of years. Some of the earliest methods of wireless communication included smoke signals and beacons. However, semaphores (optical telegraphs) probably provided the first practical form of telecommunications. Semaphores are essentially visual signaling devices that use mechanically moving arms, flags or lights to convey an alphabetic code. In 1792, French inventor Claude Chappe demonstrated the first practical semaphore system that gradually was expanded to communicate throughout France (see Figure 2.1).[1] The equipment stood on rooftops, hilltops or towers and was visible with telescopes from distances of 10 to 20 miles between stations. While a rider on horseback could take many hours to deliver a message, this new semaphore system could relay the same message in just a few minutes.

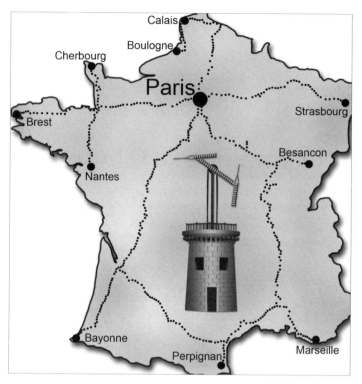

Figure 2.1 Semaphore telegraph system designed by Claude Chappe and deployed across the French landscape between 1793 and 1852. The position of the semaphore's main beam (regulator) and the outer arms (indicators) could be varied by turning handles attached to ropes in order to encode up to 196 symbols.

One of the greatest weaknesses to Chappe's semaphore system was its inability to work in fog and similar low visibility conditions. In addition, because the message needed to be manually resent from link to link, the possibility of error increased with each retransmission.

Electric telegraph

Meanwhile, efforts were underway to enable long-distance communication by means of electricity. The term *telegraphy* had already been used to describe the semaphore's operation, so this newer system was referred to as *electrical telegraphy*. The simple *electromagnet* was its central component. The electromagnet was invented in 1825 when British researcher William Sturgeon wrapped a horseshoe-shaped piece of iron with wire and passed an electric current through it. By changing the current, he created a device that was controllable by electricity.

Other inventors took notice and saw potential in using the electromagnet to send communication over long distances. By the late 1820s, electrical impulses were being sent along a wire to stimulate changes to a pointing device on the receiving end. These devices were known as *electromagnetic needle telegraphs*. In 1837, British inventors Charles Wheatstone and William Fothergill Cooke constructed a viable five-needle telegraph system that functioned by moving different combinations of needles to point to the letters of the alphabet. Like the earlier semaphore system, the message had to be read visually, but the content was carried via electricity.

American inventor Samuel F. B. Morse successfully devised the first practical electric telegraph in 1838. Morse's device was simple, consisting of nothing more than a battery, a switch, an electromagnet and some wires. Electrical current was sent through the lines. When that current was stopped and restored it was detectable on the receiving end. The earliest of Morse's designs recorded the variations as marks on strips of paper. However, it was soon discovered that the clicking sound made by the receiving device could be heard and interpreted directly by the operator. This led to the invention of a *telegraph sounder* to produce a louder, more audible noise.

Morse created a new code in which letters and numbers corresponded to sequences of short (dots) and long (dashes) electrical pulses. An operator could listen to the dots and dashes emanating from the sounder and quickly construct the individual letters into words and sentences. The Morse Code was simple, relatively easy to use, and became the universally accepted language for long-range communication on both land and sea. It continues in widespread use even today. And Morse's vision of using coded on/off signals as the basis of handling information is fundamental to computer technology.

With funding from the US Congress, the first operational telegraph line was exhibited on May 24, 1844 over a span of 40 miles between Baltimore and Washington, DC. The profound nature of this accomplishment was represented by the first message Morse sent – "What hath God wrought!" Later inventors created systems to transmit and receive multiple messages simultaneously on a single telegraph line.

Thomas Edison built the first *quadruplex telegraph* system in 1874 and successfully sent two messages in each direction simultaneously.[2]

Transatlantic telegraphy cable

The Morse telegraph and counterpart devices enabled communication over long distances – first, on land and then across oceans. The initial undersea telegraph cable was laid across the English Channel in 1850. Soon after, an American paper merchant named Cyrus Field was intrigued by the idea of a cable to span the Atlantic.[3] In the summer of 1856, he successfully linked St. John's, Newfoundland, with the rest of North America to establish a western terminus.

The transatlantic phase, connecting Canada with Ireland, proved much more challenging. Between 1857 and 1858, Fields' Atlantic Telegraph Company made several attempts to span the ocean, but all resulted in failure when the cables broke under the strain. In July 1858, the company's two ships converged in the mid-Atlantic and the cables were spliced together. On August 16, Queen Victoria sent the first telegram over the system to congratulate President James Buchanan. Unfortunately, the cable failed a few weeks later. A permanent linkage was not achieved until 1866. This ultimate success spurred numerous undersea cable construction projects throughout the world.

Telephone

Sending a message composed exclusively of dots and dashes left much to be desired. The telegraph could transmit only a few words per minute and required highly trained individuals to accurately decipher the code. The Industrial Revolution demanded a more efficient communications system. By the mid-1870s, a vast network of telegraph wires already connected cities and countries. It was natural to wonder if those lines could somehow be used to carry the human voice.

The inventors of the telephone had to figure out how to convert *acoustical energy* (sound waves) into electrical energy, carry that energy over lines, and then re-create it as sound. Acoustical energy is very different from electrical energy. Sound waves result when something physical, such as a drum or the human voice box (larynx), is disturbed and set to vibrating. The vibration of the physical object creates similar waves in the air or atmosphere. On the receiving end, human ear drums are sensitive enough to pick up these acoustical energy waves and we perceive them as sound.

For the "voice telegraph" to work, not only did sound need to be converted to electricity, but the electrical signal had to be created in such a way that it was analogous to the original human voice, with similar wave characteristics. (The term *analog* is in use today to describe this conversion of something physical to an *analogous* electrical representation.) Analog devices mimic the original physical characteristics of sound waves by varying electrical voltages and frequencies. Therefore, the device on the receiving end must be capable of re-converting the electrical representation back into its original audible form. A number of inventors set about building and improving components of what would become the telephone.

Frenchman Charles Bourseul is credited with the "original idea" in an article that appeared in the magazine *L'Illustration de Paris* in 1854. He achieved some success in the transmission of the human voice with a microphone-like device, but his receiving apparatus was a failure.[4] In 1860, German inventor Philipp Reis transmitted a message via what he called an "artificial ear."[5] His device was successful in reproducing the sound of musical tones, but failed to adequately reproduce the more complex variations of the human voice. Italian inventor Antonio Meucci, who immigrated to the United States in 1850, is often credited as the true inventor of the telephone. By 1860, Meucci had set up a primitive telephone system inside his own house to help communicate with his wife who suffered from debilitating arthritis.[6]

However, the person most associated with the invention of the telephone is Alexander Graham Bell. Bell came from a family of famous British elocutionists. (Elocution is the study of vocal production and proper speech.) This inherited interest in oral communication plus his mother Eliza's hearing loss started Bell's lifelong interest in the scientific study of sound. He eventually became a deaf education specialist at Boston University, where he increasingly focused on the ability of the human ear drum to pick up sound waves and vibrate in correspondence with them. He wondered if a man-made membrane could also cause a corresponding change in electrical current.[7]

Bell created a device with a thin metal membrane (diaphragm) surrounded by an electromagnet that received current from a battery. Sound waves falling on the diaphragm caused it to vibrate, which resulted in a conforming variation in the current. This current was transmitted over wires to a receiver where a similar diaphragm vibrated in exactly the same way as the transmitting diaphragm. On March 10, 1876, Bell spoke through his invention to his assistant Thomas A. Watson, who was located in the next room, and uttered the now famous words "Mr. Watson, come here, I want to see you."[8]

At the same time, Elisha Gray, a Western Electric Company electrician, filed a preliminary patent application for a *harmonic telegraph* – a device for transmitting vocal sounds telegraphically. Originally designed to transmit musical notes, Gray's device was then adapted to carry the human voice. Gray and Bell entered a long legal battle over patents and the right to be declared the inventor of the telephone. Ultimately, Bell's patent prevailed and his instrument led to the 1877 formation of The Bell Telephone Company.

Radio's Wireless Ancestors

The prospect of sending messages *without* wires was even more intriguing to early researchers and inventors. It began with a theory, continued with the discovery of radio waves and then culminated in a series of startling inventions.

Those who contributed to the development of wireless first had to understand the invisible waves in the air that comprised *electromagnetic energy*. Electromagnetic energy is a propagating wave in space with electric and magnetic components, possessing characteristics of both. Visible light, infrared, ultraviolet, X-rays, gamma rays, electricity, heat and radio waves are all examples of electromagnetic energy, though they differ from each other as to wavelength and frequency. For example, what we know as visible light is nothing more than waves with shorter wavelengths and higher frequencies than radio waves. Light waves can be thought of as electromagnetic energy we can see. Like the other forms of electromagnetic energy, radio waves travel very quickly, at the "speed of light" – approximately 186,000 miles (300 million meters) per second – and can flow through the air over great distances.

The road to understanding and using electromagnetic energy began in 1831 when British scientist Michael Faraday discovered that if you move a loop of wire in a magnetic field, a current will briefly flow through the wire. The effect is called *electromagnetic induction*. Next, Scottish mathematician and physicist James Clerk Maxwell described electromagnetic energy or *radiation* in a series of papers written between 1865 and 1873.[9] His theories predicted the existence of *electromagnetism*; energy that moved through space in the form of waves and traveled at the speed of light. Maxwell further suggested that messages could be sent without wires via this invisible energy. Although he did not actually prove the existence of electromagnetic energy, through reasoning and mathematics Maxwell was able to explain the relationships between electricity, magnetism and wave propagation; relationships which underpin all modern radio, television and cable communication.[10]

Maxwell provided the theoretical understanding of electromagnetic energy, but it was German physicist Heinrich Hertz who demonstrated its existence. In an 1888 experiment, he discovered that a spark from a coil of wire was able to generate a spark in a similar device across the room. The device was called a *spark-gap detector*. This transfer of energy through the air was proof of electromagnetic energy.[11] Hertz conducted further experiments to measure the length and frequency of these waves. His work was later immortalized through the practice of using the term *hertz* (abbreviated Hz) for the unit of "cycles per second" that describes the frequency of electromagnetic waves. Hertz himself did not immediately recognize the possibility of using electromagnetic waves for transmitting messages through the air, but his discovery provided the basis for the invention of wireless communication and radio.

Wireless telegraphy

Stimulated by the work of Maxwell and Hertz, inventors in a number of countries contributed to the creation of *wireless telegraphy*. A few big names often dominate the historical records, but several scientists and inventors made important contributions at approximately the same time. As early as 1868, Mahlon Loomis, a Washington, DC, dentist, transmitted a wireless message over a distance of about 14 miles between two mountains in Virginia. The system utilized kites sent aloft from two locations. The kites were covered with thin copper gauze and connected to the ground with wires. Though the system worked only sporadically, it was sufficient for Loomis to receive the first US wireless patent in 1872.[12] In 1893, Yugoslavian-born scientist Nikola Tesla demonstrated a wireless transmitter and receiver in New York City and his results were widely published.[13] However, Russian physicist Alexander Stepanovich Popov is considered by many to have been the true inventor of wireless telegraphy.[14] Popov publicly demonstrated the transmission of radio waves as early as March 1896, but never bothered to apply for a patent. To this day, March 7 is celebrated in Russia as "Radio Day" in honor of his accomplishments.

Despite Popov's contribution, the name most associated with the invention of wireless is that of Italian inventor Guglielmo Marconi. His reputation is due in large part to persistence, a good mind for business and a knack for creating excitement about the fledgling wireless technology. As a young man, Marconi had read a paper describing Hertz's experiments and became fascinated with the idea of using radio waves for communication. In 1894, he began conducting experiments on his family's estate near Bologna and within two years was able to transmit a wireless signal over a distance of 2 miles. Marconi's family was quick to see the commercial potential of Guglielmo's tinkering, but the Italian government was not interested. Fortunately, his mother, Annie Jameson, was from a wealthy British family. She arranged to set him up in England where he quickly improved his apparatus. In 1896, Marconi gave the first public demonstration of the system to military and government officials in London. By 1899, he was able to send a signal across the English Channel. A counterpart demonstration in the United States transmitted coverage of the America's Cup regatta from boats back to newspaper offices on shore.[15] Government officials were impressed and purchased Marconi wireless telegraphy systems for US military applications.

Marconi's early system utilized the spark-gap approach developed by Hertz to send Morse Code. Although the on/off aspect of the spark gap system made it useful for conveying Morse Code, there were several limitations that made it unsuitable as a practical system for long-range communication. Needed were more powerful transmitters, more sensitive receivers, signals that could be focused in just one direction (*directionalizing*), and waves that could be sent and received within a specific range of frequencies (*tuning*). The concept of tuning would become important in order for receivers to distinguish one radio signal from another and to avoid a jumble of multiple competing radio messages (*interference*).

Marconi worked on the problem of interference by building on Sir Oliver Lodge's discovery of *syntonic tuning*. In the mid-1890s, Lodge was utilizing inductors and capacitors (devices which store electrical fields) to adjust the frequency of wireless transmitters and receivers. By matching the inductors and capacitors of the transmitting and receiving antennas, Lodge was able to discriminate one set of frequencies from another.[16] In essence, the transmitter and receiver were working on the same frequency. He received a patent for syntonic tuning in 1897. However, Lodge's system lacked the power to make it a useable wireless communication device. Marconi built an improved instrument that was based on Lodge's principle in 1900 which resulted in a new method of syntonic tuning. For this work, Marconi received his famous "Four Sevens" patent (No. 7777).

Marconi's next challenge was to send a signal across the Atlantic. Many thought the task was impossible because radio waves, which tend to emanate out in a straight line, would fail to follow the curvature of the earth. Undeterred, Marconi devised a powerful 20,000 volt transmitter and placed it at a station in southwestern England. He constructed another station at St. John's Harbor in Newfoundland. On December 12, 1901, Marconi's system sent the first wireless telegraphy message between the two stations with three short bursts representing the Morse Code letter "S." How this distance was achieved, given concerns about the curvature of the earth, was unknown at the time. It wasn't until 1924 that the existence of the *ionosphere*, a portion of the upper atmosphere containing highly ionized gas, was discovered. The ionosphere bounces radio signals that would normally go straight into space back to the ground and actually increases their range. (Sometimes, inventions work long before their underlying physical principles are fully understood.)

The importance of Marconi's system was vividly illustrated in 1912 when the luxury liner *Titanic* struck an iceberg and sank. Radio operators on the *Titanic* used a Marconi wireless telegraph to send an SOS that was heard by a nearby ship, the *Carpathia*, resulting in the rescue of 717 passengers. Marconi was a shrewd businessman fully capable of exploiting such successes. He now controlled an international company with branches in Britain and the United States. His US branch later became the powerful Radio Corporation of America (RCA).

Attaining Radio

Just as Alexander Graham Bell is credited with allowing the human voice to be conveyed via wires, Canadian inventor Reginald Fessenden is recognized for his work in bringing voice to wireless. As a child, Fessenden became fascinated with telephony when he saw a demonstration of Bell's telephone. Later, as a professor in the United States, Fessenden matured into a prolific inventor holding over 500 patents.[17]

Transmitting voice and music would require a method to produce continuous waves as well as a way to impose the audio information onto these *carrier waves*. This process is known as *modulation*. Fessenden originally experimented with spark-gap transmitters; the type successfully employed by Marconi for sending Morse Code. But the spark-gap system was unable to generate the continuous waves necessary for carrying the human voice. Fessenden decided that Marconi's spark-gap approach had to be replaced by a transmission system capable of generating a continuous, sustained wave and a receiver constantly receptive to detecting these waves.[18] The problem of continual reception was solved by his 1903 invention of the *liquid barretter* (also known as an *electrolytic detector*). This device employed a thin platinum wire in a cup of nitric acid. It became an early standard for picking up wireless waves.

To create an effective transmitting device, Fessenden believed an alternating current generator, or *alternator*, could produce waves with high enough frequencies to be useful.[19] (An alternator is essentially the same thing that recharges your car battery while you drive.) In 1904, he turned to the General Electric Company, convincing them to build the needed device. Fessenden's requirement of an alternator that could produce waves with frequencies up to 100,000 cycles per second (expressed as 100 kilohertz or 100 kHz) pushed the limits of electrical technology. A young immigrant engineer at GE, Ernst F. W. Alexanderson, was given the assignment and a blunt mandate to succeed. He later recalled that, "The alternator was one of the inventions I had to make to hold my job." Fessenden and Alexanderson experienced many disagreements before the device was finished. But by 1906, Alexanderson had delivered an alternator that met Fessenden's specifications. Using the *Alexanderson Alternator*, Fessenden devised a way to vary or modulate the amplitude of the waves in order to make the variations follow the pattern of sound waves. (This is the same approach still in use for AM radio today.)

On Christmas Eve, 1906, Fessenden transmitted the world's first sound "broadcast" from his research facility in Brant Rock, Massachusetts. He talked, played music from a phonograph and performed a rendition of "O Holy Night" on his violin. The signal was heard by startled Morse telegraph operators on ships up to several hundred miles out to sea. One can only imagine their shock at hearing voices or music from a device that normally produced only dots and dashes. Fessenden's invention was referred to as *wireless telephony* because it was envisioned as a tool, much like the telephone, to allow voice communication from person to person. Unfortunately, for all his scientific brilliance, Fessenden lacked the showmanship and business sense of Marconi. His temperament often turned off potential investors, hobbling the marketing of many of his inventions.

Another important Fessenden contribution was his method for receiving continuous waves. He combined an inaudible high frequency radio wave with another steady signal in the receiver. The two frequencies created a third, whose frequency resulted from the difference of the other two (called a beat frequency) that was recognizable as sound. His resulting *heterodyne principle* brought about increased sensitivity to radio waves and enabled unprecedented clarity by eliminating much of the interference. Fessenden was unable to make a practical heterodyne receiver, but the heterodyne principle, later re-worked by American inventor Edwin Armstrong, is the basis of all modern radio reception.[20]

While the alternator produced the continuous waves Fessenden needed, it quickly became clear that this mechanical device had limitations. Alternators were noisy, large, and expensive, and they didn't produce acceptable signal quality. It was essential to move from a mechanical to an electronic apparatus more useful for transmitting speech and music.

Perhaps nothing was more important to the development of radio than the invention of the *vacuum tube* (or *thermionic valve*) – the forerunner of the modern resistor and computer chip. As early as 1883, Thomas Edison had discovered that current could be made to flow within the vacuum of a light bulb. Scientists of the time referred to this as the *Edison effect*. In 1904, John Ambrose Fleming was a scientific advisor for the British Marconi Company working on a better way to detect wireless transmissions. Based on the Edison effect, he invented the first vacuum tube. Fleming's invention was a *diode* – a vacuum tube with two electrodes (a filament and a plate) across which electrons could flow. The flow was controlled in much the same way that a faucet controls the flow of water. For that reason, it was called a *Fleming valve.*

American inventor Lee de Forest adapted the device in 1907 and created the first effective *radio tube* by inserting a control grid (a piece of bent wire or metal screen) between the filament and plate. By varying the voltage on the grid he was able to change the current flow inside the device as well. What resulted was a three electrode device (*triode*) that detected, and later amplified, radio waves. This *Audion* tube was far more sensitive than Fessenden's electrolytic detector and it changed Fleming's vacuum tube from a passive to an active component.[21]

Lee de Forest was a controversial character. He liked to call himself "the Father of Radio," but critics believe that many of his inventions were actually pioneered by others and charge that he did not understand the physical principles behind the devices he did create. Many of de Forest's business ventures resulted in claims of fraud and no fewer than 25 companies he founded went bankrupt.[22] Still, he was a prolific inventor who held more than 300 patents and it was his Audion tube that made possible the clear transmitting of speech and music. De Forest is also credited as one of the first to see the potential of radio as a medium for bringing culture to the masses. Only a year after Fessenden's first broadcast, de Forest was transmitting classical music to sailors on battleships in San Francisco Bay. He later conducted several experiments in broadcasting opera from New York's Metropolitan Opera House and in 1916, transmitted some of radio's first news reports from that same city.

Meanwhile, an electrical engineering student at Columbia University, Edwin Howard Armstrong, greatly improved the performance of the Audion in 1912 by feeding the received signal back on itself in an amplifying feedback loop. The *regenerative circuit* greatly strengthened incoming radio signals and allowed signal detection from distances previously believed impossible. Armstrong's claim to be the inventor of *regeneration* was disputed by de Forest. After a contentious 12-year patent battle the courts sided with de Forest. However, the engineering community and many historians felt that Armstrong's claim was stronger.

Though more sensitive to incoming radio signals, the Audion tube was both difficult to tune and was interference-prone. Armstrong solved the problem in 1918 with his *superheterodyne receiver*. As in Fessenden's heterodyne approach, Armstrong combined the radio wave with another steady signal in the receiver, but Armstrong was able to lower the resulting third frequency, thus making it more amenable to amplification. In fact, the signal could be amplified several thousand times; creating a device with much greater receptive ability and a tuner that could easily select stations with the turn of a knob. So effective was the device that receivers were heavily promoted in advertisements of the day as *superheterodyne radios*.

All radio at the time was transmitted by varying the height, or *amplitude*, of the waves in a process called *amplitude modulation*. This is what we know today as *AM*; but at that time, all radio transmissions used that approach. The problem with AM was that other amplitude variations in the environment, such as electrical motors and lightning, would be added to the radio signal and perceived as noise. Armstrong came upon the idea of varying the frequency of the carrier waves, instead of their amplitude. This approach, called *frequency modulation*, kept the amplitude of waves constant and made the signal immune to outside interference. Armstrong also increased the amount of bandwidth used to carry the signal by a factor of 20. The resulting *FM radio* signal was capable of providing a greater *dynamic range* – closer to the range of low pitch to high pitch sounds audible to the human ear. In addition, it was static-free. Armstrong patented FM radio in 1933; confidently predicting that FM would replace AM within five years.

Armstrong offered to sell his idea to RCA, run by his old friend David Sarnoff. But by this time, Sarnoff was more interested in developing television and preserving RCA's financial stake in AM. Sarnoff rejected the deal and a long, acrimonious dispute began. After World War II, RCA warmed to FM, but Sarnoff and Armstrong disagreed over compensation for his patents. Sarnoff was able to convince the Federal Comminications Commission (FCC) to move FM stations to a new set of frequencies on the radio dial, rendering all of Armstrong's broadcast stations and equipment useless. It was a huge financial blow. That frustration, plus the continuous legal fights over patents, left him exhausted and on January 31, 1954, the brilliant inventor took his own life. Armstrong's widow resumed the legal struggle and all of the patent lawsuits were ultimately decided in his favor. But by that time, AM had become the accepted broadcast standard. It took many more years for FM to become what Armstrong had envisioned.

The next major technological development in radio, *stereo broadcasting*, came more than 20 years after the introduction of FM and helped popularize that medium because FM radio was electronically well suited for stereo delivery. FM stations in the 1950s were desperate for any scheme that would bring them listeners and the wider bandwidth used for FM made it feasible for one station to carry two channels. In addition, stereo phonographic records were introduced in 1958, stimulating demand for stereo music over the radio. General Electric and Zenith both developed successful techniques of stereophonic transmission. In 1961, their efforts were combined to produce the modern standard for FM stereo – the *Zenith-GE Pilot Tone Multiplex* system. In order for the new stereo broadcasts to be received on existing *monaural* FM receivers, the two channels were combined (multiplexed) onto one carrier wave. Those who wanted to hear the broadcasts in stereo would require a new *stereo radio receiver* that could decode (separate) the multiplexed signal. Stereo now gave FM an edge in music broadcasting and the number of both FM stations and stereo receivers grew rapidly.

AM radio was slow to catch on to stereo. However, by the late 1970s, FM's popularity overshadowed that of AM which was now looking for a way to catch up. A number of successful AM stereo systems were devised and presented to the Federal Communications Commission. In 1982, the FCC declined to select a standard; deciding to let the marketplace determine the outcome. Motorola, Harris, and Khan-Hazeltine all fought to have their systems accepted by consumers, thus fragmenting the industry.[23] By the time the FCC belatedly adopted the Motorola standard in 1993, stations and the public had largely lost interest in stereo on the AM band.

Channels and the Electromagnetic Spectrum

What does it mean when an AM station says it's at "850 on the dial" or when an FM station calls itself "91.5"? What is the significance of a television station being at channel 3 or at channel 48? The numbers

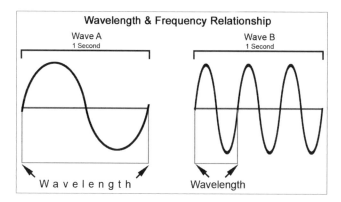

Figure 2.2 Long and short wavelengths. Wave A has a long wavelength. This means it also has a low frequency because only one complete cycle of this wave occurs in a second. Wave B is characterized by a short wavelength. This shorter length means there is a high number of waves (3 cycles or 3 Hertz) produced each second.

matter and it's all about the types of waves any given channel uses.

As we learned earlier, broadcasting waves, as well as microwaves, X-rays and even visible light, are electromagnetic waves that differ as to their wavelengths and frequencies. Figure 2.2 shows that there is a relationship between wavelength and frequency. Simply stated, long waves have low frequencies. Short waves have high frequencies.

Long wavelengths with low frequencies are generally considered to be superior waves for electronic communication. Why is this? Without delving too deeply into the physics of electromagnetic energy, it's enough to understand that long, low frequency waves behave like sound. Just like sound, these waves can go through and around objects, as well as follow the curvature of the globe – largely because of the bounce effect from the ionosphere. Instead of shooting out into space, the bounce allows waves to come back to earth. These abilities effectively increase the range of long, low frequency waves.

Short, high frequency waves behave more like light. These waves cannot penetrate most objects and are limited to *line-of-sight* transmission, meaning they do not follow the earth's curvature. Like light, they pass directly through the ionosphere and don't get the benefit of a bounce effect. For these reasons, high frequency waves have more limited range.

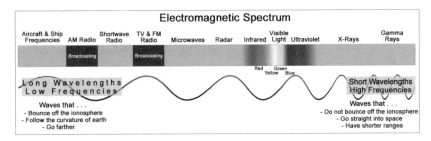

Figure 2.3 The electromagnetic spectrum chart illustrates the different characteristics of electromagnetic energy. Visible light is electromagnetic energy with shorter wavelengths and higher frequencies than the waves used for radio or TV broadcasting. X-rays and Gamma rays have extremely short wavelengths and high frequencies.

The total range of electromagnetic energy waves can be visually represented on a continuum as *electromagnetic spectrum*. Figure 2.3 illustrates that broadcasting waves are a form of electromagnetic energy characterized by longer wavelengths and lower frequencies than visible light, microwaves or X-rays.

The type of waves used for aircraft and ship communication must be capable of traveling great distances. For that reason, planes and ships use electromagnetic waves with the longest wavelengths and lowest frequencies. Those waves very effectively bounce off the ionosphere, follow the curvature of the earth and as a result, have a greater usable range.

When it comes to broadcasting (radio and TV) in the United States, the Federal Communications Commission determines where on the electromagnetic spectrum different broadcasting media are assigned. As shown in Figure 2.4, radio and television use different portions of the spectrum. AM radio, with its long, low frequency waves, can travel farther than either FM radio or television. There are differences among AM stations as well. For example, the signal of a lower frequency AM station at 580 kilohertz will have a greater *coverage area* than a higher frequency station at 1,500 kilohertz (1.5 megahertz), assuming they operate with the same amount of power.

AM radio stations are located between 535 kilohertz and 1.7 megahertz (1,700 kilohertz) and are spaced at 10 kilohertz intervals (540 kHz, 550 kHz, 560 kHz, etc.). With a *bandwidth* of 10 kilohertz for its signal, a station at 540 actually utilizes frequencies between 535 and 545 kilohertz. While their low frequency waves give AM stations a

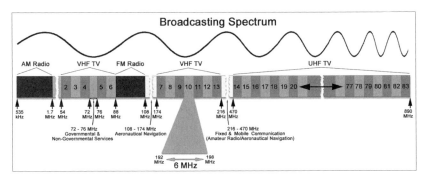

Figure 2.4 The broadcasting spectrum. The broadcasting portion of the electromagnetic spectrum shows a rather staggered approach to channel allocations. AM radio is contained in a continuous band between 535 kilohertz and 1.7 megahertz (kilo = 1 thousand; mega = 1 million). Because AM uses low frequencies, the signals from AM stations can travel great distances. FM radio appears much farther up the chart and is actually a VHF medium sandwiched between television channels 6 and 7. The spectrum for television is broken up with interruptions in the continuum for other services between channels 4 and 5, between channels 6 and 7 and between the VHF and UHF bands. A television station requires 6 megahertz of channel capacity or bandwidth, as shown with channel 10.

comparative range advantage, a station's actual coverage area depends on power, time of day and ground conductivity. The lower sound quality of AM is due to the limited bandwidth allocated for a station (only 10 kilohertz) and the fact that amplitude modulation is more susceptible to electrical interference from such things as power lines and storms.

As we learned, Edwin Armstrong became interested in FM radio (*frequency modulation*) primarily to overcome the static and noise of AM. Although it once occupied a different range of frequencies, today FM radio in the United States is situated between 88 megahertz and 108 megahertz, tucked in between the television channels we know as channels 6 and 7. FM uses a much greater amount of bandwidth than AM. Each channel is 200 kilohertz wide, but only the middle 150 kilohertz gets used for the audio signal. The outer 25 kilohertz on each side is used mainly as a buffer to avoid interference with other FM stations. FM stations can overcome the limitation of their higher frequency waves by utilizing more power for transmission. FM's superiority over AM is due to the greater bandwidth used for each station, the fact that it is less susceptible to static and its ability to reproduce sounds over a greater dynamic range.

Television stations use either VHF (Very High Frequency) or UHF (Ultra High Frequency) channels. The first television stations to come on the air in the 1940s were assigned positions in the VHF band, somewhere between channels 2 and 13. By the late 1940s there was a tremendous need for more spectrum to accommodate new stations. The FCC tried to provide space for more outlets in 1952 by allocating a new range of frequencies, UHF channels 14–83. (The upper channels 70–83 were reallocated in 1970 for pagers and mobile radio services.) However, the new UHF channels were considered inferior, because their higher assigned frequencies required much more power than VHF broadcasting to cover a comparable area.

Due to the amount of information required to transmit video, television needs an enormous amount of bandwidth for a given channel – a full 6 megahertz. For example, what we refer to as television channel 10 is really a set of waves taking up 6 megahertz of spectrum between 192 megahertz and 198 megahertz (see Figure 2.4). The number 10 is assigned by the FCC as a simple short-hand way to refer to a channel.

Keep in mind that these frequencies and channel numbers pertain to over-the-air broadcasting waves, but the majority of viewers now receive television channels via cable or a satellite service. In those cases, the channel numbers are arbitrary. A cable company can place a broadcast channel on any cable channel they desire unless the station and cable system have entered into a special agreement. For example, an over-the-air television station at channel 4 could as easily be placed on the cable system at channel 9 or channel 21.

Television Broadcasting

The idea of transmitting images by electricity pre-dated by many years the technology that made it possible. As early as 1835 there were mechanical devices capable of showing a series of printed drawings through slots and mirrors. They were known by such exotic names as Praxinoscope, Phenakistiscope and Stroboscope.[24] As with television today, these devices didn't actually produce moving images. Instead they created an *illusion of motion* through a series of still images shown in rapid succession. The illusion works because an image falling on the eye's retina is retained for a short period of time, allowing the brain to continuously connect the sequence of still images. The physiological aspect of this is called the *phi phenomenon* and the psychological component is referred to as *persistence of vision*. Research has shown that using as few as 18 images per second is sufficient to create the illusion of smooth motion.

Visions of "seeing by electricity" or "seeing by telegraph" were articulated as early as the 1870s. But, the roots of modern television date back even further to the 1817 discovery of a chemical element called *selenium*. More than 50 years later, selenium's light sensitive properties become known – quite by accident. In 1873, an Irish engineer named Willoughby Smith assigned his assistant Louis May to experiments using selenium as a resistor in underwater telegraph cables. May noticed that the resistance of selenium bars changed when hit by light from a window. After a few experiments in the lab, Smith concluded that selenium was light-sensitive and a strong conductor of electricity when exposed to light. He further discovered that the amount of current would vary depending on the amount of light hitting the selenium. The discovery of selenium's unique *photoconductive* or *photoelectric* properties led Frenchman Maurice Senlacq to propose using selenium to trace documents in 1878. He suggested that changes in electrical voltage produced by selenium could magnetically control a pencil at the receiving end of a transmission. Using the principle, British researcher Shelford Bidwell was able to send a scanned image of a silhouette in 1881. The invention was referred to as *photo-telegraphy*.

German inventor Paul Nipkow created a mechanical device capable of sending multiple images in 1884. Nipkow's device consisted of a spinning disk with a series of holes arranged in a spiral. Light reflected from the subject came through the holes striking a selenium photocell on the other side. The varying brightness on the photocell produced varying voltages which could be transmitted via electricity. On the receiving end, the process was reversed. The signal modulated a neon lamp that beamed through the holes of the receiving disk to a magnifying lens into which the viewer peered. Each rotation of the scanning disk could produce one complete image or frame. Nipkow's system essentially broke the image into *picture elements* or *pixels* arranged in horizontal lines in a process known as *scanning*. The number of lines is referred to as *resolution*. The more lines used to create an image, the higher the image's resolution. Nipkow was never able to produce a complete working model, yet his *electrical telescope*, or *Nipkow disk*, formed the basis for the first successful television systems.

Building on the concept of the Nipkow disk, two inventors created practical mechanical devices in the 1920s. In England, Scottish scientist John Logie Baird used an improved Nipkow disk to produce simple face shapes. On March 25, 1925, Baird set up the first public demonstration of television at Selfridges department store in London. The apparatus succeeded in producing silhouettes (white letters on a black card), but failed to convey a complete range of tones from dark to light. Six months later, he was able to produce an image of a ventriloquist's dummy in front of his camera in shades of gray. Baird's *Televisor* could generate 30 lines of resolution and 5 frames every second, but later versions produced up to 60 lines. In 1927, Baird transmitted a television signal over a telephone line from London to Glasgow, Scotland, and one year later, successfully sent a signal from London to New York.

At about the same time, Charles Francis Jenkins was creating a similar system in the United States. The first public demonstration of his mechanical system took place in June 1925. Like Baird's version, the first Jenkins device could produce only silhouettes, but it was the first system capable of transmitting synchronized pictures and sound. Jenkins eventually got into the manufacturing business and by the end of the 1920s, his 48–line *Radiovisors* were available to the public either pre-assembled or in kits.

By the late 1920s, it became evident that the mechanical systems developed by Baird and Jenkins had insurmountable limits. The higher scanning rates necessary to improve the picture would require an all-electronic approach. Two individuals are most often credited for subsequently inventing the *electronic television* we still use today. One was a respected scientist backed by a RCA – the other a high school student growing up on an Idaho potato farm.

The student, Philo T. Farnsworth, was a voracious consumer of scientific literature. He had read about the Nipkow disk, but became convinced that it would never produce high quality images. At age 14, reportedly while plowing the rows of a field, he came up with the idea of scanning individual lines using electrons that could be deflected back and forth with magnets inside a vacuum tube. The all-electronic system would have no moving mechanical parts and could create pictures almost instantaneously. On the receiving end, a cathode ray tube (a specialized vacuum tube with a flattened "screen" on one end) would use a beam of electrons to paint television pictures line-by-line on the back of a phosphor-coated screen. The phosphors would glow when hit by the electrons. On September 7, 1927, Farnsworth transmitted the image of a simple straight line in his San Francisco laboratory via his *Image Dissector* camera tube. It was the first electronically scanned image using electrons and it created a picture with 60 horizontal lines. Farnsworth received two patents for his television system in 1930.

But Farnsworth's contributions would become overshadowed by the work of Vladimir Zworykin, who had the backing of RCA. A Russian immigrant who arrived in the United States after World War I, Zworykin performed his early work on television between 1920 and 1929 while employed by Westinghouse. The head of RCA, fellow Russian immigrant David Sarnoff, persuaded Zworykin to join his company in 1930 and made him director of RCA's research laboratory, where he was charged with the development of television. That same year, Zworykin visited Farnsworth at his lab in San Francisco. Zworykin believed in electronic television, but was struggling to produce decent pictures with camera tubes of his own design. Farnsworth showed him how the image dissector worked, hoping RCA would offer to back his efforts.

Instead, Zworykin used Farnsworth's ideas as a guide to developing his own equipment for RCA, all the while attempting to avoid infringing on any of Farnsworth's patents. Zworykin had applied for a patent for an electronic image device called an *iconoscope* in 1923, but it was never successfully produced or tested. He built a newer version in 1931 that was very different from the 1923 original. Subsequently, RCA and Farnsworth would fight a 10-year legal battle over patents resulting in RCA paying Farnsworth a million dollars for licenses to his technology. To Zworykin's credit, he went on to create camera tubes that produced brighter images with much better picture quality than the Farnsworth image dissector. Zworykin's greatest contribution may have been his 1929 invention of the *kinescope*, a cathode-ray display which serves as the basis for television "picture" tubes still in use today.

In the 1930s, RCA began field testing broadcast television from a transmitter atop New York City's Empire State Building. In December, 1937, an industry group advised the FCC to adopt RCA's 343 line/30 frame system, but the FCC decided to wait because of opposition from other companies and a feeling that rapidly advancing technology would result in further improvements. Also in 1937, CBS announced that they would begin field testing various television systems which would be incompatible with the RCA standard. Sarnoff outmaneuvered CBS with a public demonstration of television at the 1939 New York World's Fair. From a sprawling RCA television pavilion he announced that "we are now adding sight to sound." President Franklin D. Roosevelt was introduced and became the first president to appear on "the tube."

The standards war heated up as other companies advocated their own visions for television. The 1939 RCA system was now capable of 441 lines, but DuMont proposed a 625 line system. Another company, Philco, advocated a 605-line system while Zenith felt it was still too early to adopt any standard. It became clear that these competing visions would hinder television adoption and that the government would need to impose standards to assure compatibility and growth.

On July 1, 1941, the FCC's *National Television System Committee* (NTSC) finally announced its decision. NTSC settled on the use of 6 megahertz of bandwidth for each television channel and specified that the video carrier would use amplitude modulation (AM), while the audio carrier would use frequency modulation (FM). Interlaced scanning, where the image is created in two passes – the odd lines followed by the even lines – was introduced to avoid image flicker. Interlaced scanning also allowed the television signal to more easily fit within the 6 megahertz of channel bandwidth. A 4 × 3 *aspect ratio* set the shape of the screen at 4 units wide by 3 units high and the picture would consist of 525 lines, scanned at 30 frames per second.

Much of the rest of the world adopted different standards. Besides NTSC, which is used primarily in North America and Japan, the other major world standards include *PAL* (Phase Alternating Line) and *SECAM* (Sequential Color with Memory). PAL is used primarily in Asia and most of Europe, except France. SECAM was first implemented in France, but can be found in Greece, Eastern Europe, Russia and parts of Africa. Both PAL and SECAM utilize the same aspect ratio as NTSC, but are comprised of more lines (625) and fewer frames per second (25).

With the adoption of the NTSC standards, television was ready to take off. Unfortunately, World War II intervened and video as a public medium was essentially mothballed until 1945. After the war, television adoption grew rapidly and the stage was set for the addition of color.

Less than a year after RCA demonstrated television at the 1939 World's Fair, CBS unveiled the first practical *color television* system. But there were two problems. First, the additional color information would not fit into an existing 6 megahertz channel. In fact, it would have required 16 megahertz of bandwidth – so was planned for the UHF band. Second, existing black and white sets would not be able to view the signals, requiring consumers to buy new receivers. In 1947, CBS asked the FCC to approve their color TV system. David Sarnoff saw this as a threat to RCA's large investment in black & white television, so he used his political influence to convince the FCC to wait until RCA could develop its own system that would be compatible with existing monochrome sets. Ultimately, the FCC did give approval to the CBS system in 1951, but by then the American public had made a huge investment in black & white receivers. After a few months, CBS conceded that people were simply not interested in buying new TVs just to receive color. Two years later, the FCC reversed its decision and gave approval to the NTSC color system created by RCA.

Unfortunately, RCA's method of simply adding color (*chrominance*) to the existing black & white (*luminance*) NTSC standard resulted in a serious compromise in quality, causing television engineers to label NTSC "Never Twice the Same Color." Early color sets were also expensive. It wasn't until 1965 that their cost fell below $500 and sales finally surpassed a million sets a year. That was also the same year that NBC (owned by RCA) began broadcasting most of its shows in color.[25] By 1968, color sets finally outsold black and white models.[26] Other than the rollout of stereo audio in 1984, the NTSC standard has remained largely unchanged since the 1953 acceptance of color.

Electronic Recording – Audio

Over the years, recording sound for later playback has been achieved by a variety of methods. Early devices were mechanical in nature. The *musical box*, invented in the early 1800s, consisted of a comb with teeth tuned to different pitches. The teeth were plucked by raised areas of metal on a revolving drum. Musical boxes were powered by wound springs. The *player piano*, invented in 1902, essentially played itself using small perforations on rolls of paper to activate felt-tipped wooden fingers that then pressed on the keys.[27] Early player pianos were powered by a foot pedal, but later models were operated by electric motors. These devices were marvels in their time, but neither the musical box nor player piano was capable of recording and playing back a complete representation of music or voice.

Thomas Edison's invention of the *phonograph* in 1877 was more promising and seen as a marvelous "contraption that could talk." His device was the first to successfully store and reproduce music and the human voice. Edison's early phonograph used a spinning cylinder covered with paraffin, wax or tinfoil on which a stylus made grooves. The grooves were cut to a depth that corresponded to the changes in air pressure created by the original sound waves. Emile Berliner's *gramophone*, invented in 1887, was similar, but instead used a flat disk and varied the size of the grooves side-to-side rather than cutting different depths. The flat recording disk became the industry standard for "records" well into the twentieth century.[28]

All these inventions were mechanical and the physical contact of the stylus with the grooves created friction that caused a great deal of distortion and noise. Also, the grooves were fragile and the recording medium was easily damaged. Finally, there was no way to remove, change or add material once the recording was completed.

Oberlin Smith, a mechanical engineer and industrialist from New Jersey, visited Edison's lab in 1878. He became convinced that an electrical method of magnetizing a recording medium would improve quality because it wouldn't involve direct physical contact with that medium.[29] Smith experimented with marginally successful devices, but he didn't obtain a patent and generated no fame or income from his work. His main contribution was a published 1888 description of the process in *Electrical World* magazine. The article outlined all the elements of a modern magnetic recorder, right down to the reel-to-reel transport mechanism.[30] Although there were no public displays of a working system, Smith is still credited by many as the inventor of *magnetic recording*.

Danish inventor Valdemar Poulsen picked up where Oberlin Smith left off. In 1898, Poulsen gave the first public demonstration of a working magnetic recording system – the same principle used for today's audio, video and data recording technologies. The device, called a *telegraphone*, consisted of a piano wire wrapped around a drum and a recording/playback head that moved at a constant speed across the drum via a threaded screw. The device essentially recorded sound by magnetizing the wire. The telegraphone was exhibited at the 1900 Paris Exposition where the Emperor of Austria spoke into it and recorded what is believed to be the oldest surviving magnetic recording.[31]

Magnetized wires were eventually replaced by magnetized ribbons or *magnetic tape*. Fritz Pfleumer, an Austrian inventor based in Germany, came up with the idea of embedding metal into plastic and magnetizing the metal to record sound. His 1928 invention of the *Magnetophon* was the world's first tape-based magnetic recorder. In the 1930s, German company AEG created and marketed the first practical Magnetophon *audio tape* devices and by 1943 they had developed stereo-capable recorders.

Two years later, John T. Mullin, a US Army Signal Corps engineer, heard music broadcasts from Germany and wondered how apparently recorded music sounded as good as a live performance. After the war, Mullin was in occupied Germany scouting for useful electronic devices. He managed to locate and ship two Magnetophons back to the US and on May 16, 1946, Mullin demonstrated one of the machines at a San Francisco meeting of radio engineers. Among those in the audience was Harold Lindsay, who later became a consultant for Ampex Corporation. Lindsay took precise measurements of Mullin's machine and Ampex engineers were soon working on their own audio tape recorder based on the German Magnetophon.[32] The Ampex Model 200 recorder was a massive machine by today's standards, but it was the first tape recorder built in the United States that met the quality requirements of audio professionals and broadcasters.[33]

A representative of singer Bing Crosby was also at the San Francisco demonstration. He persuaded Mullin to allow Crosby to record his radio show using the Magnetophon. Crosby disliked doing live radio and had left NBC because they wouldn't allow him to prerecord his popular program. The fledgling ABC network gained his services by allowing the practice, but Crosby was unhappy with the *disk transcription* method (recording onto LP records) then in use.[34] Initially, the Magnetophon proved an adequate substitute. But by September 1947, the Ampex Model 200 was unveiled and Crosby put in an order for 20 of them. Crosby also invested $50,000 in financially struggling Ampex; providing capital that helped the company prosper. The radio industry quickly adopted the technology and freed itself from the limitations of live broadcasts.

Electronic Recording – Video

Before there was *video tape*, the only way to record a live television show was to film it off a television monitor. These *kinescope* recordings were fuzzy and low contrast. But in the early 1950s, Lucille Ball and Desi Arnaz comprehended the potential earnings that could come from reruns of their popular *I Love Lucy* – if only a better recording vehicle was available. They convinced the studio to allow them to retain the rights to the show as long as they assumed the expense of actually filming episodes in the studio. *I Love Lucy* was shot using three film cameras in front of a live audience – a first in the history of television. Ball and Arnaz's production company, Desilu, then generated new revenue by selling the shows as reruns after they had premiered on the network. Studio and network executives began to see the value of recording programs instead of broadcasting them live. But they needed a less expensive method than film.

RCA's David Sarnoff had already directed his researchers to develop an electronic video recording device. Bing Crosby's company was working on a similar system. However, there were shortcomings with the approach used by both organizations. Video required significantly more information (bandwidth) than audio. For video, the tape had to be run through the machine at such high speeds it would go through thousands of feet of tape in just a few minutes. Such whirling machines struck fear into anyone standing near them.

Ampex, now flush with cash from its successes in audio recorders, started a video recording development program.[35] The width of the tape was greatly increased to provide a larger surface for the video information. To get the information on the tape, Ampex came up with the idea of moving the machine's heads as well as the tape. Several recording (and playback) heads were embedded in a revolving drum that moved across the tape, instead of remaining stationary as the tape traveled by. In 1956, Ampex displayed the first successful video recording and playback device at the National Association of Broadcasters convention in Chicago.[36]

The Ampex *Quadruplex* (aka Quad) system employed four video heads to record information on a 2-inch wide tape that was threaded through the machine from one reel to another (*reel-to-reel*). It utilized a *transverse scanning* approach that recorded black and white video in tracks across the width of the tape instead of in a continuous track along its length. Although the machines were expensive, Ampex received about 50 orders within days of the convention's conclusion. Two years later, the company introduced a color video tape machine. Quadruplex became an industry standard that remained in use at television stations well into the 1980s.

Ampex devised a way to edit the tape in 1958, by employing a film-like approach of physically slicing and splicing. A special videotape splicing device involved the use of a microscope, the application of an iron-particle solution and splicing tape stock to hold the sections together.[37] While cumbersome, the system was surprisingly successful. However, by the early 1960s, editing shifted to an all-electronic process in which material was copied from one tape (the playback tape) to another tape (the record tape) a shot at a time.

As successful as the Quadruplex system was, it proved unsuitable for work outside the studio. The tape was large and the machines were gigantic. In the 1960s, a number of companies experimented with a new system of recording information called *helical-scan* because the tape wound around the head drum in the shape of a helix. Unlike transverse scan that recorded information in short lines across the width of the tape, helical-scan recorded video tracks at an angle in a *slant-track* approach. The longer tracks allowed each head to record more information, while making possible narrower tape and smaller machines. These developments eventually led to the invention of the world's first small-scale, portable recording device – the Sony *U-Matic* videotape recorder.

Introduced in 1972, the U-Matic was the first recording format with the tape housed in a cassette (as opposed to an open reel-to-reel system) and it used tape that was just 3/4-inch wide. This tape format led to a revolution in television newsgathering. Prior to its invention, all news footage was acquired with film cameras. Footage had to be processed with chemicals, splice-edited, and played on the air from a film projector through a device called a *telecine* that projected the film image into a TV camera. But the 3/4-inch U-Matic cassettes could be played back on the air immediately and edited electronically. The term *electronic news gathering* (ENG), entered the TV lexicon to distinguish the new system of video from the old days of film. The designation VCR (video cassette recorder) also came about to distinguish cassette-based recording devices from the older VTRs (video tape recorders) that used open reels.

By the mid-1970s, Japanese companies such as Sony and JVC were taking the lead in developing low-cost VCRs for the mass-market that were capable of recording long-duration programs and showing movies. The Sony *Betamax* was introduced in 1975 and in that year 30,000 Betamax video recorders were sold in the United States alone.[38] A year later, the JVC *VHS* (video home system) format was introduced. A heated standards battle ensued because the Betamax and VHS systems were incompatible and the consumer had to decide between them. Experts considered Betamax technically superior, but its initially shorter recording capability (only one hour) and Sony's failure to license the technology to other manufacturers eventually led to a VHS victory. The ubiquitous VHS is still in use, although digital recording formats are gradually supplanting it.

Cable Television

There weren't many television stations in 1948 and remote areas, especially those surrounded by mountains, experienced very poor reception of those signals that were available. Appliance store owner John Walson therefore was having difficulty selling television sets at his Mahanoy City, Pennsylvania, shop some 90 miles northwest of Philadelphia. So he placed an antenna on a utility pole at the top of a nearby mountain, ran wire from the antenna down to the store, and the television sets in his window now showed the glories of good reception. People were impressed and his sales soared. But now he was under pressure to deliver those pristine signals to the homes of his customers. Walson strung more wire, added some amplifiers to boost the signal and started what was known as *community antenna television* (CATV). He agreed to hook up anyone who bought a television set from him and a year later, began charging for the service.

Similar ventures were springing up elsewhere. And department stores and apartment buildings that formerly needed a separate antenna for

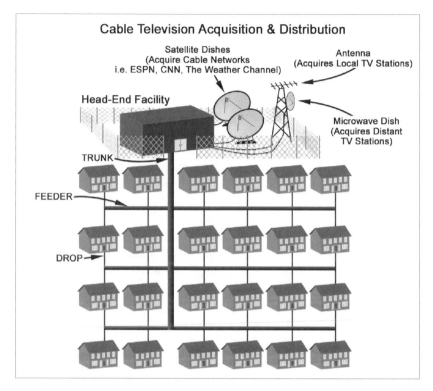

Figure 2.5 The components of a cable system

each set now utilized one master antenna (MATV), along with coaxial cable and amplifiers to distribute signals to all the receivers in the building.

The term "cable television" came to define any system that distributes television signals by means of coaxial cable or optical fibers. Figure 2.5 illustrates that a facility known as a *head-end*, collects all the signals for distribution. The head-end is usually a fenced, windowless building. Its satellite dishes pick up cable networks such as CNN, MTV, and HBO. Meanwhile, a tower with an antenna grabs over-the-air local TV stations. A microwave dish may be used to receive signals relayed from broadcast television stations within the region, but not considered local. A very large cable, called a *trunk* line, then carries the bundled signals into the community. *Feeders* split off from the trunk and bring the cable signals into individual neighborhoods. *Drops* are the cables that come into each home after branching off a feeder. Today, high-capacity fiber optic cables may be used instead of coaxial cable to make possible broadband delivery to each neighborhood. In cases where the head-end is a great distance from customers, the signal may be relayed via microwave transmission to a neighborhood box, then supplied to each house with a drop. This tree and branch system is known as a *ring-network* because the signal passes by each home whether or not it subscribes to the service.

Like over-the-air broadcasting, cables systems use electromagnetic energy. In cable the spectrum is enclosed in a wire, and though there is significant spectrum available, it is still limited. Early cable systems operated with only 200 megahertz of bandwidth. If you calculate that each television station requires 6 megahertz of bandwidth, these systems were only capable of carrying about 33 channels. ($200 \div 6 = 33.33$). Technological improvements gradually increased bandwidth capacity to 1 gigahertz (1,000 megahertz), giving systems the ability to deliver up to 166 televisions stations, or a somewhat smaller number of stations in combination with other services such as FM radio, broadband Internet, and data conveyance.

Cable operators can now provide the channels via digital means through what is called *digital cable*. Digital cable isn't necessarily the same as digital television. Digital cable is simply a method cable companies use to deliver existing stations to your home, whether or not you have a digital television set. The digital signals are decoded in the home via a set-top box and converted to an analog signal for viewing on an analog television set. Using digital compression technology, the cable system can provide high quality digital video and sound, but the service is mainly offered to provide other services such as pay-per-view programming, video-on-demand, music channels, parental controls and interactive features. Digital cable may also allow more customization of subscriber program packages.

Satellite Technology

The era of *communication satellites* (comsats) began on July 10, 1962, with the launch of AT&T's *Telstar* satellite that was used to relay television signals, multichannel telephone messages, fax and data across the

Atlantic.[39] Although it was launched into a low elliptical orbit and operated for only six months, *Telstar* was the forerunner of all modern global communications satellites.

The cable television industry was first to use communication satellites for program distribution. In the early 1970s, a nonbroadcasting entity, Time Inc.'s *Home Box Office* (HBO), was using microwave links to transmit movies and sports programming to cable systems in Pennsylvania and upstate New York.[40] However, microwave's line-of-sight requirement made it too limiting for distribution to cable systems on a national level. In 1975, HBO booked space on RCA's recently launched *Satcom 1* and became the first satellite-delivered cable channel. This opened the door to a potential national audience. A year later, the second satellite-delivered cable channel was launched with Ted Turner's Atlanta-based television station WTBS. Something previously local was now a national "superstation." By 1980, satellites contributed to an explosion of new programming services, many of them cable-only channels called *cable networks* such as CNN, C-SPAN, Showtime and ESPN.

Broadcasters didn't begin using satellites for program distribution until the late 1970s. Prior to that time, broadcast networks distributed programs to affiliates via AT&T's land-based video circuits.[41] In 1978, the Public Broadcasting Service (PBS) used a Western Union *Westar* satellite to deliver programs to local PBS affiliates, becoming the first broadcast network to use a satellite. In 1984, NBC became the first commercial television network to employ them and the other commercial networks soon followed.

By the mid-1980s, program *syndicators* – the people who sell network reruns and first-run programs such as game and talk shows – had changed from a system of physically shipping film or tape copies of programs to stations to using satellite distribution. At about the same time, local television stations started dispatching *satellite news gathering* (SNG) vehicles to cover stories. Stations still used line-of-sight microwave-equipped trucks for local live-remote news reports, but when the distance was too great or the location too remote, SNG vehicles were deployed.

A large number of backyard satellite dishes also sprang up in the 1980s Some of these households paid for the programs they received, but many illegally "pirated" the signals intended for cable systems. This piracy growth led the cable industry to scramble their transmissions, making them available only to authorized cable systems. But the public was clearly interested in *direct-to-home broadcasting* (DTH) via satellites and a new industry was soon born to legally deliver it.

Direct-broadcast-satellite (DBS) services thus arrived in the 1990s. Although the multi-state footprint of DBS "birds" made it impractical to provide the local stations that cable systems carried, DBS companies could offer more overall channels than cable and with higher picture quality. The quality and capacity of DBS, along with accessibility of DBS in remote areas, led to tremendous growth.

From the standpoint of technology, communication satellites are merely relay platforms in the sky. Video, audio and data from earth are *uplinked* from a dish antenna and received by one of a satellite's *transponders*. The transponder amplifies the signal, shifts it to another frequency and then retransmits it (*downlinks*) back to earth where it is picked up by a receiving dish antenna. Each satellite carries multiple transponders, allowing it to receive and send a large number of channels simultaneously. Companies lease transponder space on a satellite to distribute their programs.

To avoid having to hit a moving target, a satellite is "parked" in a geosynchronous orbit at an altitude of about 22,240 miles above the earth's equator. At this altitude and location, a satellite's orbital speed equals the earth's period of rotation so it stays in a fixed position relative to the ground below. From this vantage point, a satellite can transmit to an unlimited number of ground receivers within its *footprint* or coverage area.

Just as AM and FM radio use different portions of the spectrum, satellites can vary as to the range of frequencies they use. Early satellites used an area of spectrum between 3.7 and 4.2 gigahertz for downlinks and were called *C-band*. Satellites transponders using this frequency range required a great deal of power and the receiving dishes, known as television receive-only (TVRO) installations, were very large – often 6 to 12 feet in diameter. In the 1980s, a new generation of satellites called *Ku-band* became popular. Ku-band signals used higher frequency waves, between 10.9 and 12.7 gigahertz, and required less power. The Ku-band also permitted the use of much smaller receiving dishes, making it advantageous for satellite news gathering and direct broadcasting satellite applications. Both C-band and Ku-band are in use today. The satellite industry is utilizing digital compression technology to send more television signals in less bandwidth, thus greatly increasing the number of signals each satellite can deliver.

Analog vs. Digital

Most of the previously discussed electronic media inventions, such as AM, FM and NTSC television, are *analog* devices. This means that electrical waves are created in such a way that they mimic the original human voice (sound waves) or the properties of light reflected from objects shot by a video camera. In other words, these continuous electrical waves vary so as to be "analogous" to the properties of the transmitted subject. However, there are drawbacks to using electricity in this way. The process introduces noise, static and distortion. These problems become worse as the signal is transferred or copied from one analog device to another. This can be seen if you've ever copied a VHS video tape, then used the copy to make another copy. The result is a degraded image and reduced sound quality – something called *generational loss*.

Conversely, a *digital* system is built upon numbers, usually binary numbers such as 0 and 1. These numbers are coded to represent something else. The concept is not new. Even a smoke signal is a binary digital system in the sense that a puff of smoke represents "on" (1) and no smoke represents "off" (0). The puffs of smoke don't mean anything unless we assign meaning to them, so in that sense they are coded. The Morse Code used a non-binary, but digital system of five coded states – dot, dash, short gap (to separate letters), medium gap (to separate words) and long gap (to separate sentences).

These days, the term digital usually refers to the binary number system used by computers. Unlike their analog counterparts, digital devices don't suffer generational loss. Digital copying is called *cloning* because as long as all the numbers are correctly transferred from one digital storage medium to another, the copy is an exact duplicate of the original.

Once computers became powerful enough, inventors and engineers determined that audio and video information could be stored via *digital* means. Because 0s and 1s are discrete values, rather than a continuous spectrum of energy such as electrical waves, the challenge was to use enough numbers to adequately represent the waves of sound and light.

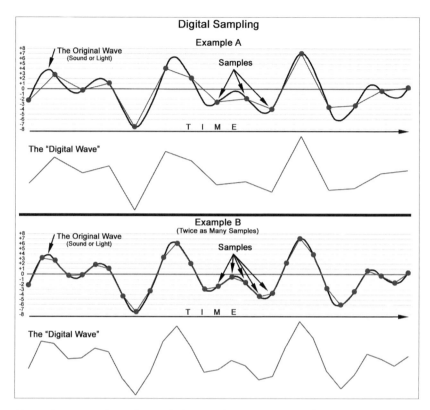

Figure 2.6 Digital sampling. Which "digital wave" most resembles the original wave? In example A, using just a few samples (the dots) results in a very crude representation of the original wave. Example B uses twice as many samples – resulting in a smoother wave that more closely approximates the look of the original. If the original wave was audio, the sampling process in example B would result in much higher sound quality than example A. (CD-quality audio actually uses 44,100 samples every second.) If the original wave was video, the video in example B would be sharper and clearer with more accurate reproduction of colors.

Computers don't create a complete representation of a wave. Rather, they *sample* the wave and store numbers to describe it. For that reason, a wave's digital replication is only as accurate as the number of samples used to represent it. It will never be as complete a picture as an analog representation which is continuous. Figure 2.6 shows how digital sampling is improved by increasing the number of sampling points used to re-construct the wave. Taking more samples results in better quality, but it also means the computer has to process and store more information.

For this reason, it took many years before computers were powerful enough to handle the enormous quantities of information required for video.

The ubiquitous compact disc (CD) was the first consumer device to play back digitally recorded audio. Some of the first digital devices used in radio production were *digital audio tape* (DAT) machines.

On the video side, the first digital device was a *time-base corrector*. Invented in 1972 by Consolidated Video Systems, it was used to convert analog VCR playback to a digital signal for the purposes of stabilizing images produced by helical scan VCRs. In 1975, the British company Quantel released a *digital framestore* to allow picture-in-picture insets for coverage of the 1976 Montreal Olympics. At around the same time, *digital video effects* (DVE) units came on the scene. Now, video producers could make shots spin, rotate, tumble and scale. The first DVE from a company called Vital Industries was marketed as a *Squee-Zoom*, because that's what it did to video images.

Later, computers became fast enough to allow video producers to *digitize* material and edit programs without the loss normally associated with analog (tape-to-tape) editing systems. The editing process was called *digital non-linear editing* because it permitted shots to be edited in any order and easily re-arranged on the computer. Earlier analog tape-based editing required building a program in a linear fashion. The order of shots was difficult to change and new shots were nearly impossible to insert without re-editing much of the program. Today almost all video editing for broadcasting, and even movies, is accomplished on a computer via digital non-linear editing.

Digital Television

The story of *digital television*'s development in the United States involves political, economic and technological sub-plots. By the 1980s, NTSC television was more than 40 years old, but vastly improved television systems were being formulated in Europe and Japan. Japanese companies were already the dominant manufacturers of consumer electronics and the US feared losing out in the high-tech race for a new and better television system. The Japan Broadcasting Corporation, NHK, had begun development of a high-definition (HDTV) system in the 1970s. After years of work, NHK initiated a satellite-delivered service called *Hi-Vision* in 1989. Hi-Vision was designed to provide 1,125 lines (1,035 lines for the picture itself) with the goal of creating an image that would rival the quality of 35 millimeter film. While the system utilized some digital processing, it was essentially analog. The Japanese system was demonstrated in the United States in 1987 and the impressive display caused some broadcasters to consider Hi-Vision's adoption.

At the same time, broadcasters in the United States saw high-definition television deployment as a way to hang on to the valuable broadcast channels under threat of re-assignment to mobile communication companies.[42] Not all the VHF and UHF channels were being used and companies such as Motorola wanted to acquire some of the UHF channels for two-way radio services. Broadcasters claimed they could instead utilize the unoccupied channels to offer high-definition television. Political protectionism and industry self-interest led to the determination that a US system should and could be developed; a system that would surpass the Japanese by being all-digital. It might also preserve the essence of local broadcasting by using traditional land-based antennas, instead of a satellite delivery system like Japan's.

In 1987, the FCC announced it favored using the UHF channels for high-definition television and established the *Advisory Committee on Advanced Television Service* (ACATS) to promote competition among companies and to evaluate *advanced television* (ATV) systems.[43] However, most of the early proposals were for analog systems. The technological hurdles to digital television were huge because digital television is essentially a computer and computers of the era were unequal to the challenge. Computer processing speed, memory and storage were simply too limited for the requirements of all-digital television. Also, the large amount of data required for digital television seemed too great for the bandwidth of a 6 megahertz broadcast channel. Would two, or perhaps three, channels be needed to broadcast a single digital signal?

In 1990, ACATS made three crucial decisions: (1) the new system should be high-definition, rather than simply an improved or *enhanced-definition* system; (2) it would *simulcast* high-definition programs on a VHF or UHF channel separate from the conventional NTSC program transmission; (3) the new signal could not exceed 6 megahertz of bandwidth.

General Instrument was the first to demonstrate that a digital high-definition system was possible and that it could fit within a

6 megahertz television channel. One of the company's engineers, Woo Paik, went home for a week of uninterrupted thinking and came up with a way to squeeze a high-definition television signal inside a 6 megahertz channel.[44] *Interframe compression* was the key. Paik's system did away with the redundancy of television frames by transmitting only the changes that occurred from one frame to the next. For example, when a newscaster speaks and the camera doesn't move, the only area that needs to be transmitted is the anchor's mouth. The unchanging set and backdrop don't need "repeating" in every frame. Within six months, three competing digital systems and one analog system were proposed by other companies.

Beginning in the summer of 1991, the systems were scrutinized in a testing facility called the *Advanced Television Testing Center* (ATTC) with the goal of selecting a winner that ACATS could recommend to the FCC. Unfortunately, nearly two years of testing produced no consensus. Each system had pros and cons and instead of choosing one, the FCC suggested the companies ought to merge their efforts and come up with a single composite HDTV standard. Therefore, on May 23, 1993, the digital television Grand Alliance was formed to combine the best features and capabilities of systems designed by former competitors the Massachusetts Institute of Technology, Zenith-AT&T, and the Philips-Thomson-Sarnoff consortium.

An early debate within the Grand Alliance pitted broadcasters against the computer industry. Broadcasters favored the interlaced (alternate line) method of scanning used by NTSC. The computer industry preferred the *progressive scanning* method used by computers. (Progressive scanning creates the lines of an image sequentially from top to bottom without skipping any lines.) Broadcasters and television set manufacturers argued that progressive equipment would be more costly, but other parties felt that the benefits of merging television and computer technology would justify a progressive scanning approach.

During the next two years, the system developed by the Grand Alliance was refined and tested. In November 1995, ACATS unanimously recommended adoption of the *Advanced Television Systems Committee* (ATSC) *Digital Television* (DTV) standard. It received formal FCC approval on Christmas Eve, 1996.

Where the NTSC standard adopted in 1941 specified a single standard for scanning (interlaced), lines of resolution (525 − 480 for the picture), aspect ratio (4 × 3) and frame rate (30 frames per second),

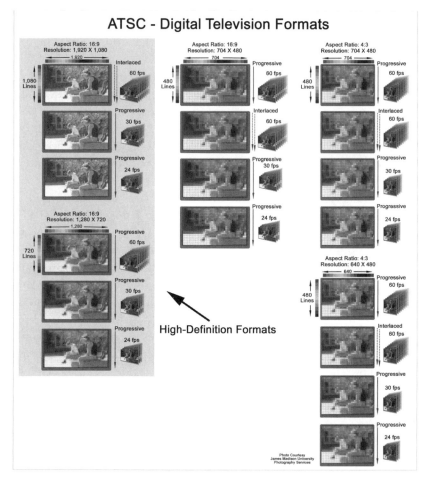

Figure 2.7   ATSC digital television formats. The 18 ATSC formats vary as to screen shape or aspect ratio, resolution (the number of lines making up the picture), scanning method and frame rate. Only the six formats on the left are considered high-definition, which was the original goal in creating an advanced television system.

the new DTV protocol was a compendium of many standards. Digital Television could be produced, transmitted and displayed with any of 18 different combinations of scanning, resolution, aspect ratio and frame rate. The picture might be scanned as either interlaced or progressive, it could use 480 lines, 720 lines or 1,080 lines, it could be either 4 × 3 or 16 × 9 aspect ratio and could generate 24, 30 or 60 frames per second. Because digital television sets are essentially computers, these receivers would be able to decode what was being transmitted and automatically show the program in the proper format. Only the 720 and 1,080 line formats were considered *high-definition*. Video transmitted at 480 lines was called *standard-definition* because its resolution was comparable to NTSC. The ATSC created a table of formats, called *Table 3*, which laid out the 18 different formats for DTV (see Figure 2.7).

In reality, the most common formats are 1,080i (i = interlaced), 720p (p = progressive) and 480p. The 1,080-line and 720-line formats utilize a 16 × 9 aspect ratio – a widescreen shape more like motion pictures. However, even 480-line signals can be produced and transmitted at 16 × 9.

The FCC also gave each broadcaster a temporary second channel to use during digital television's phase-in. The 6 megahertz of spectrum on the new digital channel could carry approximately 19 megabits-per-second (Mbps) of data. Thanks to compression, 19 megabits can deliver one 1,080-line high-definition signal or multiple lower quality channels – what are referred to as *sub-channels*. In other words, broadcasters could decide to transmit a single program stream (in HDTV), or they could send out two, three or even four separate non-HDTV programs all in the space of one existing channel. Such *multicasting* wasn't originally envisioned and Congress believed they had authorized these new digital channels expressly to carry HDTV. But broadcasters figured out that transmitting four non-HDTV services rather than a single HDTV one would allow them to become multichannel providers – rather like mini cable systems. More services would lead to more advertising, thus generating increased revenue. Any leftover bandwidth could be used to deliver data as well. The advent of business *datacasting* suggested a completely new revenue stream for broadcasters who were historically reliant solely on advertising.

The ATSC standard also specified that the signals would be transmitted using *MPEG-2 compression*, *Dolby digital surround sound* and *Program and System Information Protocol* (PSIP) data. The PSIP data make it easier for viewers to find where a station's new digital channel exists. It essentially maps the new channel to the existing NTSC channel. For example, the viewer accustomed to picking up a CBS station on channel 4 will still get it on channel 4, even though it may be sent over the airwaves on channel 47. If the channel has four subchannels, the receiver identifies them as 4-1, 4-2, 4-3 and 4-4.

At the turn of the century, the world of DTV remained a complex place for consumers. DTV broadcasts were hard to find, only a select few cable systems were carrying the signals and satellite services often required premium subscriptions for DTV content. Those wishing to receive DTV over-the-air had to consider the *digital cliff* effect that takes into account how far one lives from the broadcasting tower. With analog transmissions, as the signal strength is reduced, the picture gradually becomes noisier. With digital transmission, the quality of the picture remains unchanged until suddenly the picture completely vanishes. One viewer in a neighborhood may receive a beautiful high-definition signal, but a block or two away the viewer may get nothing at all – having fallen off the digital cliff.

Chapter Rewind

Some of the earliest methods of communicating over a distance involved smoke signals, beacons and semaphores (optical telegraphs). A semaphore system developed by Claude Chappe spanned much of France in the late 1700s, reducing the time to transmit a message from hours to just a few minutes. Samuel F. B. Morse invented the first practical electric telegraph in 1838. Perhaps Morse's greatest contribution was his code. The Morse Code's system of coded on/off signals is still in use and is the basis of handling information in today's computer technology.

Efforts to expand the reach of electronic communication led to the development of undersea cables. American paper merchant Cyrus Field successfully laid the first transatlantic cable in 1858. By the mid-1870s, a vast network of telegraph wires connected cities and countries around the world.

The next challenge was to carry the human voice via wires. Alexander Graham Bell's voice telegraph (telephone) was among the first devices to convert sound to electricity and re-convert electricity back to sound on the receiving end. The process of converting sound waves to analogous electrical signals provides the origin of the term "analog."

Wireless development began with the discovery of radio waves or electromagnetic energy. First theorized by James Clerk Maxwell and then proven to exist by Heinrich Hertz, electromagnetic energy was successfully utilized by Guglielmo Marconi in the late 1890s in his wireless telegraph. The wireless telegraph connected continents and permitted communication between ships at sea.

The next steps in developing wireless communication involved efforts to carry the human voice and music. Reginald Fessenden convinced General Electric to build him an alternator, a device capable of producing the continuous radio waves necessary to carry sound waves. On Christmas Eve, 1906, Fessenden transmitted the first broadcast from his research facility in Brant Rock, Massachusetts. The vacuum tube was invented in 1904 and by 1907, Lee de Forest adapted the device to create the first effective radio tube called an Audion. Later, Edwin Howard Armstrong improved the Audion's reception by adding a generative circuit. With his 1918 superheterodyne receiver, Armstrong solved tuning and interference problems and went on to invent and develop an improved system of radio known as frequency modulation or FM.

The invention of television goes back to the 1817 discovery of the light-sensitive chemical element selenium. Selenium was first used by Paul Nipkow in 1884 in his spinning disk which was capable of converting light into electricity. During the 1920s, the Nipkow disk was refined in England by John Logie Baird and in the United States by Charles Francis Jenkins. Their devices produced crude images, but were successful in transmitting television signals from one location to another.

These mechanical devices gave way to all-electronic systems of television invented and developed by Philo Farnsworth and Vladimir Zworykin. In 1939, RCA created a public sensation with the exhibition of television at the New York World's Fair. Two years later, the federal government adopted a standard for television in the United States known as NTSC. Color was added in 1953 and stereo audio came along in 1984.

Electronic methods of recording audio were boosted by the postwar acquisition of a Magnetophon, a magnetic tape recording device developed in Germany and brought to the United States following World War II. Based on the device, Ampex developed the first audio tape recorder used by professionals and broadcasters. The company went on to develop the first workable video tape recording system in 1956 – its Ampex Quadruplex.

Cable and satellite technology provided new methods of audio/ video delivery in the 1950s and 1960s. The 1970s and 1980s witnessed the development of digital devices for storing and manipulating audio and video. A vision for an all-digital, high-definition system of television to replace the aging NTSC led to intensive research in the late 1980s and early 1990s. In 1996, the FCC adopted a new digital television (DTV) system to be phased in as a replacement for NTSC.

SELF-INTERROGATION

1   What were the greatest weaknesses to communicating via Claude Chappe's "optical telegraph"?
2   How did the ionosphere play a role in Marconi's successful effort to send a wireless signal across the Atlantic?
3   Reginald Fessenden created what "first" in 1906 from Brant Rock, Massachusetts, and why was it referred to as wireless telephony?
4   Why was FM radio especially well suited for the development of stereo radio broadcasting?
5   Why do radio stations with lower frequencies have longer ranges than stations with higher frequencies?
6   Why were the television technologies developed by John Logie Baird and Charles Francis Jenkins said to have insurmountable limits?
7   What 1939 event publicly launched television? Who sponsored it and why?
8   Describe the competition between CBS and RCA to create a color television standard. Why did the RCA system eventually take hold?
9   Why was the 3/4-inch U-Matic video recording system so important and what part of the television industry did it revolutionize?
10   In what ways did satellite technology contribute to changes in program distribution for the cable and broadcast television industries?
11   What is sampling rate and how does increasing the number of samples improve digital recording?
12   What does it mean when a broadcaster uses their digital television channel for multicasting?

NOTES

1   Chappe Telegraph System, "Optical Telegraphy: The Chappe Telegraph Systems"; available from http://people.deas.harvard.edu/~jones/cscie129/ images/history/chappe.html; accessed 5 September 2005.

2   IEEE Virtual Museum, "The Quadruplex Telegraph"; available from http://www.ieee-virtual-museum.org/collection/tech.php?taid=&id=2345885&lid=1; accessed 10 November 2005.

3   Bern Dibner, *The Atlantic Cable* (Cambridge, MA: Burndy Library, 1959), 7.

4   Adventures in CyberSound, "Charles Bourseul: 1829–1912"; available from http://www.acmi.net.au/AIC/BOURSEUL_BIO.html; accessed 11 November 2005.

5   Johann Philipp Reis; available from http://chem.ch.huji.ac.il/~eugeniik/history/reis.html; accessed 11 November 2005.

6   The Garibaldi–Meucci Museum, "Antonio Meucci"; available from http://www.garibaldimeuccimuseum.org/antoniomeucci.html; accessed 11 November 2005.

7   John Brooks, *Telephone: The First Hundred Years* (New York: Harper & Row, Publishers, 1976), 38.

8   American Treasurers of the Library of Congress, "Mr. Watson – Come Here!"; available from http://www.loc.gov/exhibits/treasures/trr002.html; accessed 18 November 2005.

9   Edgar E. Willis and Henry B. Aldridge, *Television, Cable and Radio: A Communications Approach* (Englewood Cliffs, NJ: Prentice-Hall, 1992), 21.

10 John Bray, *The Communications Miracle: The Telecommunication Pioneers from Morse to the Information Superhighway* (New York: Plenum Press, 1995), 21.

11 Marvin Smith, *Radio, TV and Cable* (New York: Holt, Rinehart and Winston, 1985), 57.

12 Kazimierz Siwak and Debra McKeown, *Ultra-Wideband Radio Technology* (Chichester: John Wiley & Sons, 2004), 9.

13 A. Michael Noll, *Principles of Modern Communication Technology* (Norwood, MA: Artech House Inc., 2001), 102.

14 M. Radovsky, *Alexander Popov: Inventor of Radio* (Honolulu, HI: University Press of the Pacific, 2001), 50.

15 Gian Carlo Corazza, "Marconi's History," *Proceedings of the IEEE*, vol. 86, no. 7, 1309, July 1998.

16 IEEE Virtual Museum, "Oliver Joseph Lodge"; available from http://www.ieee-virtual-museum.org/collection/people.php?taid=&id=1234730&lid=1; accessed 21 November 2005.

17 The Public Forum Institute, "Reginald Fessenden, Wireless Radio Transmission"; available from http://www.publicforuminstitute.org/nde/global/innovators/fessenden.htm; accessed 30 November 2005.

18 Susan J. Douglas, *Inventing American Broadcasting* (Baltimore, MD: The Johns Hopkins University Press, 1987), 45.

19 Lewis Coe, *Wireless Radio: A Brief History* (Jefferson, NC: McFarland & Company, 1996), 10.

20   John S. Belrose, "Fessenden and Marconi: Their Differing Technologies and Transatlantic Experiments During the First Decade of This Century"; available from http://www.ieee.ca/millennium/radio/radio_differences. html; accessed 1 December 2005.

21   Jerry C. Whitaker, *Power Vacuum Tubes Handbook* (Boca Raton, FL: CRC Press LLC, 1999), 2.

22   Harold Evans, *They Made America* (New York: Little, Brown and Company, 2004), 223.

23   Philip Hoff, *Consumer Electronics for Engineers* (Cambridge: Cambridge University Press, 1998), 42.

24   David Mellor, *A Sound Person's Guide to Video* (Woburn, MA: Focal Press, 2000), 147.

25   Roger F. Fidler, *Mediamorphosis: Understanding New Media* (Thousand Oaks, CA: Pine Forge Press, 1997), 98.

26   Coe, *Wireless Radio*, 124.

27   Mark Coleman, *Playback: From Victrola to MP3, 100 Years of Music, Machines and Money* (Cambridge, MA: Da Capo Press, 2003), 4.

28   Jim Dawson and Steve Propes, *45 RPM: The History, Heroes and Villains of a Pop Music Revolution* (San Francisco, CA: Backbeat Books, 2003), 7.

29   Andre Millard, *America on Record: A History of Recorded Sound* (New York: Cambridge University Press, 1995), 33.

30   Eric D. Daniel, C. Denis Mee and Mark H. Clark, *Magnetic Recording: The First 100 Years* (New York: IEEE Press, 1999), 7.

31   Joel Levy, *Really Useful: The Origins of Everyday Things* (Willowdale, Canada: Firefly Books, Ltd., 2002), 161.

32   Albert Abramson, *The History of Television, 1942 to 2000* (Jefferson, NC: McFarland & Company, Inc., 2003), 21.

33   Daniel, *Magnetic Recording*, 83.

34   Brian Winston, *Media Technology and Society: A History from the Telegraph to the Internet* (London: Routledge, 1999), 266.

35   Frederick Wasser, *Veni, Vidi, Video: The Hollywood Empire and the VCR* (Austin, TX: University of Texas Press, 1997), 59.

36   Paul Martin Lester, *Visual Communication: Images with Messages* (Belmont, CA: Thomson Wadsworth, 2006), 332.

37   Gary H. Anderson, *Video Editing and Post-Production: A Professional Guide* (Woburn, MA: Focal Press, 1999), 4.

38   Matt Haig, *Brand Failures: The Truth about the 100 Biggest Branding Mistakes of All Time* (London: Kogan Page Limited, 2003), 26.

39   Kevin S. Forsyth, "Delta: The Ultimate Thor," in Roger D. Launius and Dennis R. Jenkins (eds) *To Reach the High Frontier: A History of U.S. Launch Vehicles* (Lexington, KY: The University of Kentucky Press, 2002), 116.

40 Megan Gwynne Mullen, *The Rise of Cable Programming in the United States: Revolution or Evolution* (Austin, TX: University of Texas Press, 2003), 7.

41 Andrew F. Inglis, *Satellite Technology: An Introduction* (Woburn, MA: Focal Press, 1997), 22.

42 Joel Brinkley, *Defining Vision: The Battle for the Future of Television* (New York: Harcourt Brace & Company, 1997), 4.

43 Hugh R. Slotten, *Radio and Television Regulation* (Baltimore, MD: The Johns Hopkins University Press, 2000), 241.

44 Joan Van Tassel, *Digital TV Over Broadband: Harvesting Bandwidth* (Woburn, MA: Focal Press, 2001), 89.

CHAPTER 3

Content Chronicles

Peter B. Orlik

As was previously discussed, technological evolution provided the hardware and the avenue via which the electronic media were and are able to reach audiences of every size and type. But technological capability means little if the content this technology carries does not resonate with media consumers. How such content evolved is this chapter's subject.

Broadcasting: Stumbling upon a Business

In Chapter 2 we learned that early "wireless telephone" experimentation was mainly concerned with the field-testing and improvement of equipment in order that valuable patent rights could be staked out. Initial activities were the innovations of applied scientists rather than mass communicators. Program content considerations were incidental to the fact that this content was being successfully transmitted via its sender's invention. But when Dr. Frank Conrad and other engineers discovered that private citizens were actually interested in the subject matter they had randomly chosen for their test transmissions, the possibilities for a whole new industry began to emerge.

A Westinghouse Electric Company researcher, Conrad was operating experimental station 8XK out of his Pittsburgh-area garage as a means of field-testing prototype wireless components. But when the Joseph Horne department store began advertising pre-built wireless sets in September, 1920, so that more people could listen to Conrad's makeshift programming, Westinghouse executive Harry Davis sensed a whole new business opportunity. 8XK increased its power to 100 watts and secured a license from the Department of Commerce as KDKA. This marked the shift from "wireless" as a point-to-point activity between private stations to "broadcasting" radiated to the general public.

The shift in focus from industrial hardware production to consumer program creation came quickly, but not immediately. Historian Frederick Lewis Allen recalls the 1921 radio scene:

> Dr. Van Etten of Pittsburgh permitted the services at Calvary Church to be broadcast, the University of Wisconsin gave radio concerts, and politicians spouted into the strange instruments and wondered if anybody was really listening. Yet when Dempsey fought Carpentier in July, 1921, and three men at ringside told the story of the slaughter into telephone transmitters to be relayed by air to eighty points throughout the country, their enterprise was reported in an obscure corner of the *New York Times* as an achievement in "wireless telephony"; and when the Unknown Soldier was buried at Arlington Cemetery the following November, crowds packed into Madison Square Garden in New York and the Auditorium in San Francisco to hear the speeches issued from huge amplifiers, and few in those crowds had any idea that soon they could hear all the orations they wanted without stirring from the easy-chair in the living room.[1]

By the following year, this "easy-chair listening" began to become a reality. Consumers could now buy pre-assembled radios from an increasing number of retailers and scores of new stations came on the air to promote the newspaper, store, or equipment manufacturer that owned them.

Even more important, when WEAF's (New York) sale of airtime to a Jackson Heights real estate developer demonstrated that a station could make money *in its own right*, the programmers and salespeople pushed the inventors and engineers out of the industry's driver's seat. Within a few months, three new trade magazines (*Radio World, Radio Dealer*, and *Radio Broadcast*) had been launched to serve the emerging mass medium and several universities put stations on the air or were making plans to offer courses by radio.[2] When heavy November snows in the Rocky Mountains crippled telegraph lines, 1922 election returns got through on the radio. The print media trade magazine *Editor and Publisher* lauded this election service in an editorial entitled "Radio's Increasing Value to the Public" – but the US print establishment would soon come to perceive radio as more competitive threat than public servant.[3] Though radio was simultaneously blossoming in other countries around the globe, any economic threat to their commercial press was largely blunted as stations came under the control of state-owned and networked noncommercial corporations. Unfortunately for the publishers, that was not to be the US pattern.

US networking, or *chain broadcasting* as it was often called, began in 1923 with AT&T's permanent linking of six outlets. Informally known as the Red network because of the red ink that AT&T engineers used to map out their interconnections, this operation expanded to a coast-to-coast linkup in little more than a year. Its programming would eventually become the cornerstone of the NBC empire.

Such mushrooming commercial activities were not always viewed in a positive light, however. Many people thought that the airing of commercials was an unconscionable theft of the public airwaves for private gain, and newspaper publishers increasingly worried whether such practices might eventually lure advertisers away from their pages. Even the clergy were becoming concerned. Several church services were regularly broadcast by 1923 and some clerics protested the practice because it seemed to be diminishing actual church attendance. One bishop even declared that sinners couldn't be converted by radio because the apparatus could not fully convey the necessary magnetism of a clergyman's personality.[4] Despite such fears, the public as a whole remained wildly enthusiastic about the new medium. As *Scientific American* commented at the beginning of 1924: "Looked upon as a fad in the beginning, radio broadcasting has now entrenched itself pretty firmly in the routine of American life. This is due to the commendable effort of the broadcasters who, during the past twelve months, have been steadily improving their programs."[5]

With the 1924 presidential campaign, radio demonstrated both its penetration and its broad appeal. Millions of listeners tuned in as the Democratic Party's convention struggled ballot after ballot to pick their presidential candidate. People never before involved in the political process now found it to be a highly charged enterprise and eagerly followed developments from the privacy of their homes. This intense interest did not go unnoticed by the newspaper industry. Dominated by its large-publisher members, and fearing this new electronic competitor, the Associated Press's board of directors voted to withhold all AP news copy from stations.

Meanwhile, RCA vice-president David Sarnoff was encouraging the new business to stand on its own through the self-sufficiency that the sale of airtime could bring. "Broadcasters must be able to pay their own way in order to stabilize the industry," Sarnoff proclaimed.[6] Many stations did just that by starting to peddle time blocks to advertisers. Sponsor names began appearing in show titles. On February 22, 1924, the N.W. Ayer advertising agency inaugurated the first advertiser-sponsored entertainment series when it launched battery-maker National Carbon Company's *Eveready Hour*, a dramatic anthology.[7] Insurance firms started to associate themselves with health and exercise programs, and home recipe shows specified the brand name of the flour that should be sifted. *Product placement* had begun. By 1925, major market stations were charging as much as $500 an hour for program sponsorship,[8] and the true business of broadcasting had emerged. In fact, the new National Association of Broadcasters trade group formally decided to use the term *broadcasting* to reference its enterprise because the previous term, *radiocasting*, had been grabbed by the Associated Manufacturers of Electrical Supplies.[9]

The growth and solidification of network activities further helped to anchor the new mass medium. When AT&T withdrew from broadcasting in 1926 in a deal to protect its domination of land lines and long-distance activities, it transferred its stations and the associated Red network to a group that soon came to be dominated by the Radio Corporation of America (RCA). RCA then formed the National Broadcasting Company (NBC) to manage the Red network and its own much smaller Blue network. Needing an audio logo, NBC appropriated a three-chime signature using the musical notes G, E, and C in honor of the General Electric Company, NBC's largest stockholder other than RCA.[10]

Soon after the announcement of NBC's birth, a combination of talent agents and independent stations who feared they would be marginalized by the dual-net NBC formed United Independent Broadcasters as a competing chain in 1927. The Columbia Phonograph Company subsequently became interested, and the organization was renamed the Columbia Phonograph Broadcasting System. But the network had hardly been launched when Columbia became dismayed by the mounting costs and lack of sponsor support. It pulled out of the project, leaving only its name behind. Fortunately for the struggling company, a young cigar manufacturing executive had done a little radio advertising and liked the results. William S. Paley's family owned the Congress Cigar Company, which had been one of Columbia Phonograph Broadcasting's first clients. When the 26-year-old executive found that radio commercials for his La Palina cigars doubled the brand's sales,[11] the power of this new content vehicle became clear. For about $500,000, Paley's family purchased a majority of the chain's stock and he remained active in directing what would be known as CBS until his death in 1990.

At the time the Columbia Broadcasting System was born, there were already well over 500 US radio stations on the air, reaching a weekly audience of 23 million through almost 6 million radio sets. From the beginning, sports and music constituted major program elements. Harold Arlin, the industry's first full-time announcer and former Westinghouse electrical engineer, hosted the first broadcast of a major league baseball game over KDKA in 1921 and country music's *Grand Old Opry* took to the air via Nashville's WSM in 1925.[12]

Early US broadcasting's biggest breakthrough, however, came in the comedy genre. Beginning in 1925 under the title *Sam 'n' Henry*, the show starred Freeman Gosden and Charles Correll, two white entertainers who specialized in black dialect. Three years later, the Chicago program moved from WGN to competitor WMAQ as *Amos 'n' Andy* where it was performed live while disc recordings were made for shipment to more than two dozen cooperating stations. This was one of the first examples of program *syndication* via which a show is sold to individual stations without regard to their network affiliation. In 1929, *Amos 'n' Andy*'s wide appeal led NBC to obtain this 10-minute serial comedy on behalf of Pepsodent toothpaste.

One of the first national telephone surveys of the radio audience found that more than half of the people called were listening to *Amos 'n' Andy*.[13] This astonishing hit made 7 P.M. nightly radio listening a national past-time. At the time, very few people thought the concept of two white men performing black caricatures to be racist or even of questionable taste. In a society that was still segregated in its day-to-day activities, *Amos 'n' Andy* was enjoyed as easily absorbed humor. It relied on the naïveté of the roles portrayed to quickly convey plot and laughter to an audience oblivious to the show's distorted assumptions. Like the vaudeville stage from which it and so much early radio comedy had sprung, *Amos 'n' Andy* sacrificed truth for selective simplicity in order to appeal to the largest listenership possible. The show proved how popular radio could be. In hindsight, it also demonstrated that broadcast popularity and social consciousness do not automatically go hand in hand.

Radio's Gilt-Edged Years

As the world entered the 1930s and the economic depression that became that decade's hallmark, it was feared that this financial calamity would stifle the further growth of radio. As it turned out, the medium hardly broke stride. Listeners gravitated toward their ornately encased radio set (or "wireless" in the British parlance) as a form of essentially free entertainment (see Figure 3.1). Where countries that had adopted noncommercial systems relied on annual receiving set licenses to pay the bills, Americans let sponsors pick up the tab. Advertisers saw radio as the most efficient way to stay in touch with consumers. And performers found it vital to the sustenance of their careers and bank accounts as their audiences had less money to spend on out-of-home entertainment. True, the average cost of an hour of radio time on a major US station dropped to $310 in 1930,[14] but this was still a healthy figure when compared with the free-falling prices in other industries. To stimulate their own sales, newspapers began to accept advertising from broadcasters, and the "radio page" became an important source of program announcements, reviews, and paid layouts promoting individual shows.[15]

Radio was also becoming a political force. Astute politicians around the world came to understand that radio was something more than a diversion. Even before his first "Fireside Chat" on March 12, 1933, new US President Franklin Delano Roosevelt demonstrated his grasp of broadcasting. Rather than stiffly approach the microphone as though it were a public address system, Roosevelt talked through it conversationally, as if each listener were being individually approached. The power of FDR's voice soothingly masked the fact that his legs were

Figure 3.1   A 1930 state-of-the-art Atwater Kent floor model radio. Source: Lautman Photo 77377, the Smithsonian Institution.

crippled from polio. Radio thereby constituted one of the most potent tools of his lengthy presidency. Meanwhile, the chief executive's affinity for broadcasting provided the medium with increased stature. In short, Roosevelt and radio proved an ideal, mutually supportive match.

Radio listenership continued to increase even as the full impact of the Depression caused radio revenues to decline. Comedy and variety programs like the *Eddie Cantor* and (Rudy) *Vallee Varieties* shows on NBC-Red, *Amos 'n' Andy* on NBC-Blue, and *Burns and Allen* on CBS provided listeners with a momentary escape from the troubles of the times. Serials and soap operas such as *Little Orphan Annie* (NBC-Blue) and *Just Plain Bill* (CBS) continued to reinforce daily listening habits. Coverage of college and professional sporting events increased while team proprietors debated whether such broadcasts helped or harmed stadium attendance. And radio news readings and high profile on-the-spot coverage treated the public to quick and captivating bursts of information.

Major newspaper interests awoke to the fact that radio had become more competitor than ally. In 1933, the Associated Press membership, dominated by old-line publishers, voted to ban network use of AP news and to limit local use of their wire copy to brief bulletins that credited the local AP-member newspaper. The American Newspaper Publishers Association decided that radio program schedules were no longer news but would be printed only if paid for as advertising. An awkward pact between the two media was concluded early the next year with the establishing of the Press-Radio Bureau. Through this agency, the wire services agreed to provide stations and networks with two five-minute summaries per day (one that must be broadcast after 9:30 a.m. and the other after 9:00 p.m. so as not to beat the release of morning and afternoon papers). These newscasts had to be *sustaining* (unsponsored) programming and each ended with the sentence: "For further details, consult your local newspaper."[16]

The networks dismantled their fledgling news-gathering operations and agree to abide by these restrictions. But a number of stations balked. News exchanges like the Transradio Press Service, the Continental Radio News Service, and the Radio News Association were created to serve the journalistic aspirations of client stations. Before long, they surpassed the Press-Radio Bureau in the scope and quality of their activities, and the networks found themselves at a competitive disadvantage as the public's appetite for instantaneous news grew.

Earlier in 1934, a new network came on the scene with the organization of the Mutual Broadcasting System. Distinct in its operating philosophy from the other networks, MBS owned no stations and was primarily concerned with securing advertising for its member outlets, who were each completely free to make their own programming decisions. Pioneered by it four charter members: WGN (Chicago), WOR (New York), WLW (Cincinnati) and WXYZ (Detroit), Mutual gradually added more and more independents to its fold. Beginning with the widely popular *The Lone Ranger* from WXYZ (see Figure 3.2), MBS was able to serve its members' commonality of programming interests and advertising needs without compromising the right of each station to operate as it saw fit.

Figure 3.2   A mid-1940s rehearsal of *The Lone Ranger*, the series that helped launch MBS programming a decade earlier. The Ranger, played by Brace Beemer, is gesturing with a balding Tonto (John Todd) at his side. Source: Courtesy of "Wyxie Wonderland" by Dick Osgood.

Despite the continuing Depression, by 1935 the radio industry was riding a rapidly rising wave of profits and popularity. A CBS survey indicated that seven out of ten US homes possessed receivers. Revenues from the sale of advertising time increased 20 percent over the previous year. This was seen to be radio's Golden Age. But in many ways, argues business historian John McDonough,

> the golden age of radio was really the golden age of the ad agency . . . The main creative function of an agency serving a national account was creating programs, not commercials. The theory was the message was less important than its environment; that listeners would be grateful to sponsors for providing free entertainment and buy their products out of gratitude; and that over time product and program would become one by association.[17]

More often than not, from the aforementioned *Eveready Hour* in 1924 onward, this seemed to be exactly what happened.

Almost 600 broadcast stations were on the air in 1935 (all AM, of course), and a few insiders thought viable FM service might not be far away. Even important segments of the hostile newspaper industry became convinced of broadcasting's financial potential. By the middle of the year, 114 stations were owned by publishers.[18]

Serials, comedies, musical varieties, and quiz shows were massively popular, but so too were serious drama, education and music offerings. CBS's *Columbia Workshop* aired outstanding aural theater by writers like Archibald MacLeish. NBC produced four Eugene O'Neill plays, and both network companies competed as to who could mount the finest radio adaptations of Shakespeare. Musically, CBS broadcast New York Philharmonic concerts, and NBC countered by securing the services of renowned conductor Arturo Toscanini to lead its new in-house ensemble. Under Toscanini's baton, the NBC Symphony Orchestra premiered on Christmas night, 1937, for what would turn out to be a 17-year run. Even purely educational programming became a network battleground. To trump the CBS *American School of the Air* that brought 6 million children programs on geography, history, English, music and drama,[19] NBC launched *University of the Air*. Clearly, these were glorious times for cultured dial-twisters who were getting programming the equal of any being produced by other countries' state-run systems.

Still, the golden age did have its tarnishing elements. For one thing, local stations increasingly injected their own advertisements into sponsored programs they received from the networks. These spot commercials (or just "spots" for short) brought in additional revenue but were not welcomed by the national clients who had paid handsomely for exclusive program sponsorships. Ford Motor Company declared that the spots were "unfair to the sponsor and to the public."[20] Nevertheless, the networks were leery of totally outlawing the practice for fear of losing important affiliate stations.

Programming itself came in for increasingly negative criticism as the tastes of urban listeners and broadcasting executives sometimes clashed with the sensitivities of rural consumers. With four dominant networks (NBC-Red, NBC-Blue, CBS, and MBS) now serving virtually the entire United States through their local station "chains," regional taste divergences inevitably caused friction. A 1937 example was the NBC Sunday presentation of an "Adam and Eve" skit on the early evening *Chase and Sanborn Hour* in which the voluptuous Mae West played Eve. Even though Ms. West's prodigious dimensions were not discernible over the radio, her racy reputation was well known. The casting of this actress in a skit based on biblical characters raised a storm of criticism that found its way onto the floor of the US House of Representatives via a complaint by Congressman Donald L. O'Toole.[21] Although the congressman's protest resulted in no formal action, his attack mirrored concerns that would resurface anytime broadcasting's ready access to the home and children was thought to have been abused.

Simultaneously, the unsettling effect of radio content on even adult psyches was proving a source of dismay as radio news brought the specter of real catastrophe and radio drama evoked the clever devices of science fiction apocalypse. When WLS announcer Herb Morrison, on a routine assignment at Lakehurst, New Jersey, recorded his on-the-spot account of the German dirigible *Hindenburg*'s explosion, the NBC audience was soon an observer to the conflagration. A few months later, as Adolph Hitler's armies marched into Austria and dismembered Czechoslovakia, CBS, NBC and MBS brought the events and the German dictator's speeches directly to the living room via recorded and later, live transatlantic shortwave relays. A new breed of authentic radio journalists such as Edward R. Murrow, H.V. Kaltenborn, William L. Shirer, and Fulton Lewis, Jr., introduced US listeners to a fascinating and foreboding world they could no longer ignore. These news actualities may have made fictionalized news seem more credible as well. Orson Welles's *The War of the Worlds* broadcast on Halloween Eve 1938 was only a dramatization of an H.G. Wells science fiction classic. But radio had now conditioned people to receive bad news almost instantaneously. Welles's *Mercury Theater on the Air* depiction of a Martian invasion caused thousands of listeners to panic and led CBS to drop the use of simulated news bulletins within entertainment programming.

In 1940, with important real war news breaking every minute, the Associated Press gave up the fight to keep radio out of the news business. It agreed to accept stations as members and shut down its information-stifling Press-Radio Bureau. With Edward R. Murrow's chilling actualities of bombs raining on London and the William Kierker (NBC) and William L. Shirer (CBS) broadcasts of France's surrender to Hitler, radio journalism acquired increased stature and dramatic appeal. Buoyed by a *Fortune* magazine survey of a few months earlier that "70 percent of Americans relied on the radio for news and 58 percent thought it was more accurate than that supplied by the press,"[22] broadcast correspondents and commentators became prime partners in the dissemination of vital information to the American public.

The 1941 Japanese attack on Pearl Harbor further heightened the importance of radio news and impacted all aspects of the broadcast industry. Military priorities meant the diversion of electronic parts, equipment, and personnel to more essential uses than commercial broadcasting, and radio had to make do for the duration. Entertainment programming turned more and more to war-related themes. Popular songs like "Praise the Lord and Pass the Ammunition" shared the airwaves with the war bond drives, *Little Orphan Annie*'s serialized run-ins with diabolical Nazi spies, and soap operas in which characters anxiously discussed the fate of their servicemen husbands and sons. In a special radio address on December 9, 1941, that reached an unprecedented 90 million listeners, President Roosevelt set the tone and the ground rules for broadcasting in the years to come:

> To all newspapers and radio stations – all those who reach the eyes and ears of the American people – I say this. You have a most grave responsibility to the nation now and for the duration of this war. If you feel our government is not disclosing enough of the truth, you have every right to say so. But, in the absence of all the facts as revealed by official sources, you have no right in the ethics of patriotism to deal out unconfirmed reports in such a way as to make people believe they are gospel truth.[23]

With former CBS news commentator Elmer Davis at its helm, the Office of War Information (OWI) worked to ensure that Americans at home and abroad received as much news as possible consistent with national security. Broadcasters complied fully with the content restrictions imposed in Pearl Harbor's wake. In addition, a number of government/industry cooperative productions were aired, including *Command Performance* and *The Army Hour*. The 11 private shortwave stations taken over by the OWI helped make such programs available to members of the armed services abroad as a means of countering the propaganda of Tokyo Rose and Berlin's Axis Sally. Out of such activities grew the Armed Forces Radio Service.

At home, meanwhile, war priorities served to give broadcasting a competitive edge over the print media because paper rationing forced both newspapers and magazines to downsize. Advertisers who could no longer purchase adequate page space bought more airtime, and radio advertising revenues reached new heights. Even classical drama and symphony orchestra presentations were given a consequent boost, as sponsors clamored to support any available programming, no matter how highbrow, if it would keep their names in the public's memory.

As the war dragged on, the country and the radio industry were forced to come to grips with another threat – the increasing polarization and hatred spawned by more obvious racial discrimination. A product of its times, the US military was uncompromisingly segregated; a fact that Axis Sally's Berlin broadcasts constantly exploited with fabricated reports of the cowardice of black units. To counter these lies, War Department officials joined with some of US radio's finest writers to produce programs that accurately reflected the contributions of black Americans to the war effort. An episode of William Robson's *Man Behind the Gun* dramatized the exploits of the Coast Guard cutter *U.S.S. Campbell*'s all-black crew, which sank a half-dozen enemy submarines. Soap operas introduced heroic African Americans into their plot lines, and Mutual's *Fighting Men* and *Men O' War* on CBS provided vehicles for black servicemen to relate their training and combat experiences.

Such efforts came none too soon. In 1943, a Detroit race riot killed almost three dozen people. CBS's response was the preparation and airing of *Open Letter on Race Hatred*, a half-hour dramatization that *Time* magazine called "the most eloquent and outspoken program in radio history."[24] Produced by William Robson as a replacement for that week's episode of his *Man Behind the Gun*, the show was first sent down the network's closed-circuit lines for affiliate preview. Several southern stations declared they would not air the program. Meanwhile, the CBS affiliate's owner in Detroit, who had taken out full-page newspaper ads promoting *Open Letter*, was furious when he learned the presentation did not take an anti-black stand. Nonetheless, the program went out throughout the rest of the country with an announcer introduction that framed the show's courageous stand and prophetic warning:

> Dear fellow Americans. We ask you to spend thirty minutes with us facing quietly, and without passion or prejudice, a danger which threatens all of us. A danger so great, that if it is not met and conquered now, even though we win this war, we shall be defeated in victory. And the peace that follows will for us be a horror of chaos, lawlessness and bloodshed. This danger is – race hatred.[25]

Ultimately, almost 100 CBS affiliates carried the broadcast, and radio demonstrated its potential, if not its unanimous dedication, to serve the public interest, regardless of the consequences.

Post-War Adaptations

With the end of World War II in late 1945, the last vestiges of wartime censorship and restrictions on *ad lib* programs (suspected of facilitating coded communication among enemy agents) were removed. The largely experimental FM activity was moved to a higher and more spacious portion of the spectrum in preparation for full roll-out, and broadcasters eagerly looked toward a predicted postwar consumer boom. New quiz shows like Mutual's *Break the Bank* came on the air, and existing ones sweetened their prizes. By late 1946, the number of AM stations had jumped from approximately 1,000 to more than 1,500 and almost 400 FM outlets were authorized by the FCC. Radio's prospects, it seemed, couldn't be better.

The future, however, held bleaker surprises. FM was almost immediately dubbed the "Free Music" medium because scarce receivers and scarcer advertiser support inhibited more expensive and varied live programming. Even on mainstream AM, an advertiser slump was materializing. Plenty of sponsor money had been available during the war – not only because of print media space limitations but also because high taxes on excessive wartime profits coerced companies to plow these profits into advertising rather than pay them to the government. With these taxes now lifted, and paper readily available for unrestricted newspaper and magazine publication, the competition for limited sponsor dollars intensified.

And then there was television. On the surface, network radio seemed as vibrant and competitive as ever. But through skilled application of the tax laws, CBS Chairman William Paley was luring major radio stars such as Jack Benny, Edgar Bergen, and Red Skelton from NBC and Bing Crosby from ABC (the former Blue network which NBC had divested in 1943). Paley helped these stars to become independent corporations, so they could effectively tax shelter much of their income. The CBS chairman was not interested in continuing the radio careers of these personalities. Instead, he clearly saw their television potential and the "pre-packaged" success they could bring to CBS's video ventures. Soon, the other network companies also moved to exploit their radio talent resources in the new television industry. Even before the end of the decade, key radio programs were being seen rather than just heard. Along with CBS's *The Goldbergs,* ABC's *Stop the Music* and *The Lone Ranger* were among the first programs to make what would soon be a wholesale migration of national namesakes from radio to television.

The 1950s dawned unpleasantly for the sound medium. Network radio revenues declined significantly as more and more sponsors followed their stars to television. The onset of the Korean War brought renewed interest in radio news, but General Douglas MacArthur, supreme commander of the US armed forces, imposed extensive censorship on reports from the war zone. For a time, big network radio shows continued as NBC premiered a new all-star variety program called *The Big Show* featuring Tallulah Bankhead. But other radio personalities like Jean Muir, Henry Morgan and Martin Gabel were finding their names on network blacklists as the US Congress continued its Communist-seeking witch hunts.

Much of the escalating drop in radio listenership was occurring in the evening as former listeners were becoming viewers of television's night-time fare – which had often been appropriated from radio. The radio networks began to turn more and more airtime back to their affiliates and the affiliates generally responded with the new "disc jockey" shows. Thus, the networks' longstanding policy and practice of airing only live music were replaced by local stations' preference for far cheaper recorded programming. Station owners Todd Storz in Omaha and Gordon McLendon in Dallas derived significant success with a format called "Top Forty" in which the most popular records of the day were endlessly replayed by youth-appealing deejays who communicated in short bursts between tunes. Simultaneously, disc jockey Alan Freed dominated the Cleveland airwaves by spinning a pulsating blend of rhythm-and-blues, country and gospel music he dubbed "rock and roll" (borrowing a phrase from a 1922 blues song).[26] With its almost hypnotic appeal to teenagers, rock and roll soon became a national phenomenon propelled by local stations. It gave local radio a new franchise with young people while their parents were deserting the medium for comfortable seats in front of "the tube."

Between 1947 and 1955, the percentage of radio stations with network affiliations dropped from 97 percent to 30 percent – even as FM rolled out and many more AM stations were born. The consequent decline of sponsor-packaged network programs and rise of local deejay shows allowed the radio commercial to fully come into its own. Previously, when advertising agencies owned network shows, they focused their creative resources more on the programs than on the commercials they inserted into them. But now, their commercials had to be inserted as "spots" into thousands of local environments, competing with thousands of other advertising messages. Consequently, observes commercial historian John McDonough:

> The passing of radio's golden age ironically paved the way for the golden age of the radio commercial. Bob and Ray made the adventures of Bert and Henry as effective for Piel's Beer as Myrt and Marge had once been for Wrigley gum. And Dick Orkin and Bert Berdis dramatized the urgency of reading *Time* magazine in award-winning spots.[27]

Just like local broadcasting, the network radio that remained came to be funded by "spots" rather than program sponsorships. Networks also increasingly turned to short news and musical segments that would augment local stations' disc jockey shows. A pioneering example was NBC President Sylvester "Pat" Weaver's creation of *Monitor*, a potpourri of interesting modules strung together from 8 a.m. Saturday until midnight on Sunday. *Monitor* served both listener lifestyles and affiliate scheduling needs by providing a wealth of easily digestible features designed to enhance a variety of listener and local station use patterns.

By the late 1950s, radio had stabilized by embracing its new local and music-driven role. Buoyed by rock and roll's pied piper impact on youth, local radio was posting impressive revenue gains as advertisers flocked to reach the young "baby boomer" consumer market in its members' formative years. The "payola" scandals (in which disc jockeys were caught taking bribes from promoters to push certain records) gave the industry a temporary scare. But tightened station policies – and high-penalty "anti-payola" amendments to the federal Communications Act – largely pushed the issue from sight by the early 1960s.

In November, 1960, the final bastions of old network radio fell as the last four long-running soap operas (all on CBS) left the air for good. *Young Dr. Malone*, *Right to Happiness*, *The Second Mrs. Burton*, and *Ma Perkins* (which had been a fixture since 1933), tied up their plot lines and faded away as their sponsors abandoned them in favor of their developing video soaps. The radio populace became more and more fragmented. Local stations began to more carefully target specific audience segments so as to more efficiently appeal to both national and local advertisers. Phone opinion programs came on the scene as a way to increase the involvement and loyalty of target listeners. People were encouraged to call in and chat with the show host while the rest of the audience eavesdropped on the conversation. Soon, these offerings evolved into entire talk-radio blocks that exploited the intimacy of the increasingly portable radio medium.

Initially, FM stations had a distinct portability disadvantage as few car radios possessed FM capability. In fact, the lack of FM receivers in general had proved a huge problem for the new audio medium even as local AM prospered. The situation changed rapidly in the mid-1960s however. First, the Federal Communications Commission approved a 1964 regulation that limited to 50 percent the amount of time FM stations in major cities could duplicate the programming of a commonly owned AM outlet. This finally forced owners to make FM programming something different. Beginning in 1966 on WOR-FM (New York) and given west coast prominence over San Francisco's KMPX(FM), a new format known as "Progressive" or "Free Form" rejected both shouting Top Forty deejays and the formal voices found on old-line "adult" stations. In their place, Progressive enlisted laid-back, conversational communicators who featured album cuts excluded from conventional playlists.[28] Forerunners of AOR (album-oriented rock) and similarly formatted outlets, Progressive stations abandoned the rigid structures of conventional programming to give the youth of the late 1960s an antiestablishment radio service they could call their own. Thus attracted to FM, most of these young listeners grew to forget that AM ("antique modulation") ever existed.

A few commercial stations tried to be different by avoiding advertising completely. They sought, instead, to secure direct listener support for their specialty programs through on-air begging. Usually offering such off-beat material as jazz, folk songs, operas, and Broadway soundtracks, such stations pleaded with their audience to send in money as a fair return for not having to listen to commercials. Unfortunately, these on-air appeals were often much more intrusive, and much less entertaining, than real advertisements would have been. (The pledge drives on today's noncommercial stations sometimes resemble these unrelenting "commercial subscription" pleas.)

In fact, commercial subscription stations were soon pushed aside when a reorganized US "public broadcasting" came to the fore and usurped much of the subscription stations' content. The 1967 passage of the Public Broadcasting Act set up the Corporation for Public Broadcasting (CPB) as a vehicle to support the expansion of noncommercial radio and television. Four years later, National Public Radio (NPR) was created as a distributive agent for the larger non-commercial stations to help them produce and share high-quality cultural and special-appeal programming. Through a series of gradually increasing facility, schedule, and personnel requirements, NPR membership came to embrace and assist both established and new stations (usually operating on the "reserved noncommercial" portion of the FM band) to become major voices in their localities. CPB and other federal funding sources dovetailed their requirements with those for NPR membership so that a system of strong and professionally managed stations emerged. However, many existing public school and student-operated college stations could not meet CPB/NPR resource standards and were permanently bypassed in this upgrading process.

Radio's Second Half-Century

Radio entered the 1970s with heightened attention to format differentiation and an FM band poised for prominence. To distinguish FM's superior sound from the lower fidelity AM, programmers developed new music-intensive genres. A format success in one market was quickly cloned in another as format doctors (program consultants) like Bill Drake developed and sold concepts around the country. Drakes's "Solid Gold" and Bonneville's "Easy Listening" formats helped make profit centers from what had been money-losing FM outlets. Meanwhile, some maverick FMs created an anti-format known as "underground." An extension of the 1960s Progressive FM phenomenon, underground radio mirrored the social turbulence of its times, providing disenchanted youth with their own broadcast voice. As Dick Kiernan, one of its chief practitioners recalled, underground outlets ventured far beyond their progressive rock roots. They were also

> playing Wagner; they were playing Czechoslovakian folk music; they were playing just an incredible gamut of music. And they were talking a lot. They were playing 20 and 30-minute segments from philosophers like Marcuse. In some instances, it was extremely creative, but no one can be extremely creative 24 hours a day 7 days a week. The bursts of creativity were few and far between but they were brilliant when they occurred.[29]

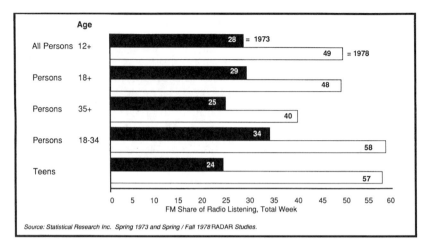

Figure 3.3   FM's surge to dominance: 1973 and 1978 shares of US radio listening compared

Through the influence of master programmer Lee Abrams, underground would eventually mellow out in the post-Vietnam War era into more formula-based album-oriented rock (AOR) patterns with narrowed playlists and greater predictability.

As FM became more and more popular, (see Figure 3.3), key old-line AM stations turned to talk formats that minimized AM's fidelity disadvantage. Beginning late in the 1960's with Gordon McLendon's XTRA (a Tijuana, Mexico, station serving Southern California), the *all-news* format caught on in several major markets. All-news capitalized on radio's portability and a mobile society's need to get quick bullets of information at any time of the day or night. Although expensive to program due to its high personnel costs, all-news could turn significant profits through its appeal to upscale listeners and the advertisers seeking to reach them.

By the early 1980s, new "occasional use" networks like NBC's The Source, and CBS's RadioRadio arose to service the FM boom with music features attractive to younger listeners. New program syndicators also came on the scene with both short-form (individual features) and long-form (complete format) offerings. As satellite delivery of these services became readily available, such activity multiplied because stations and suppliers were now liberated from the limitations and costs of landline distribution. Gradually, the distinction between radio syndicators and networks melted away. In fact, at mid-decade, syndicator

Westwood One purchased both the Mutual Broadcasting System and NBC Radio. That two historic networks could be swallowed up by one syndicator showed how much the audio environment had changed and consolidated. Both AM and FM outlets could now get whatever type of material they wanted, in virtually any length, from a wide range of program services without regard to a single network "affiliation."

Network/syndicator merging was paralleled by 1990s station ownership consolidation. The 1996 Telecommunication Act practically eliminated radio ownership limits, paving the way for the growth of giant groups like Clear Channel, which subsequently accumulated more than 1,200 radio stations. Owning up to eight outlets in a single market, Clear Channel and other mega-groups could dominate certain format clusters, and rely on satellite-delivered programming to fill the time on multiple stations around the country. *Voicetracking* deejays were now employed to serve several outlets simultaneously, bringing higher quality talent to the local airwaves but also undercutting truly local programming.

The advent of direct-to-consumer satellite radio in the new century's first decade, however, may well reinvigorate local radio. Satellite program services Sirius and XM can digitally send well over a hundred program channels to subscribers' homes, cars and trucks throughout the land. They can free high-profile and controversial talents like Howard Stern from the content restrictions inherent in the operation of FCC-licensed stations. But digital satellite's wide-footprint character also means it cannot deliver truly local content. To exploit this limitation, even outlets owned by the mega-groups are reconsidering how important localism may be to their future well-being, In addition, stations are countering the satellite threat by streaming their radio signals and other programming over their websites to reach new listeners, upgrading to high-definition (HD) service with sound quality that rivals that of the satellites and above all, reminding listeners that a radio station's signal is free.[30]

Video Enters the Living Room

NBC's April, 1939, telecast of the New York World's Fair opening signaled the US start of regularly scheduled video programming intended for public consumption. In the days that followed, the handful of

New Yorkers with sets could watch studio snippets of plays, puppet shows, comedy and singing acts, and remote sporting events captured by NBC's cumbersome two-bus mobile unit. One of these events was also groundbreaking from an advertising standpoint. During the first major league baseball telecast (a game between the Brooklyn Dodgers and Cincinnati Reds), sportscaster Red Barber delivered live pitches for three Dodger radio sponsors who had been given TV time as a "value added" bonus. Mr. Barber established the TV spot when he donned a Sonoco hat to plug a can of motor oil, held up a bar of a Procter and Gamble brand soap, and sliced a banana into a bowl of Wheaties.[31]

These early telecasts by NBC and other electronics pioneers were purely experimental in nature. The FCC did not grant regular commercial operating authority until July, 1941, when an upgraded 525-line picture was ready for roll-out. Five months later, television was essentially mothballed as components and resources were diverted to the development of war-related materials. Commercial television did not re-emerge until 1946. By that time, the few pre-war sets had been rendered obsolete and the building of a consumer industry had to start essentially from scratch with only nine operating stations on the air. It was not until 1947 that true television *broadcasting* first started to hit its stride, much in the same way that radio had begun to blossom in 1921.

Before 1947 was over, advertisers were lining up to buy airtime on the 12 operating stations, the public was waiting cash-in-hand for the sets that were now rolling off the production lines, and enterprising tavern owners in cities served by television were finding that a receiver over the bar was a potent method for attracting new clientele. Bab-O (a powdered cleanser) became one of television's pioneer sponsors when female viewers were offered a green plastic replica of a good-luck beetle-shaped brooch once worn by the ancient Egyptian queen Hatshepsut. (Bab-O did not reveal that this beetle was also an ancient fertility symbol; perhaps further energizing the Baby Boom.) Soon ABC was flooded with envelopes containing the required quarter and Bab-O label. This type of "hook" or premium commercial became a TV advertising staple because it also helped sponsors to gauge the size and composition of the new medium's audiences.

The DuMont network temporarily took the lead in program development, premiering *Mary Kay and Johnny*, television's first situation comedy, after having introduced the medium's first soap opera,

Faraway Hill, the previous year.[32] Together with riveting telecasts by DuMont's competition of a World Series and a heavyweight championship boxing bout, these visual offerings grabbed broad consumer attention. As *Fortune* magazine observed, "Public interest in TV as a gadget is so great that there is no question but that receiving sets in quantity can be sold. . . . Buyers will be willing to take their chances on the quality of the programs."[33]

This gamble started to pay off more broadly for American consumers in 1948. The price of a standard television set had dropped to about $300; a month's salary that the average working person was eager to pay. Once the set was in the house, viewers were treated to the excitement of the 1948 political conventions and the 15-minute nightly network newscasts read by Douglas Edwards on CBS and John Cameron Swayze on NBC whose *Camel News Caravan* was wholly sponsored by a cigarette company. ABC lacked the resources to be competitive in news or television programming generally, but did manage to put its New York flagship station on the air during the year.

Meanwhile, DuMont (the only company that had carried on with extensive program activities during the war years) linked up its stations in New York, Pittsburgh and Washington. Founded and controlled by electronics expert Allen B. DuMont, the independent company manufactured cameras, transmitters, and associated station equipment as well as a line of television receivers. Even though it would ultimately lack the programming clout to compete with the "Big Three," DuMont pioneered television networking in a number of ways and also introduced significant innovations in consumer sets, including a "giant" 30-inch oval picture tube. Some consumers had to tear out doorways to get its cabinet into the house. "Who wants to look at the world through a knot hole?" asked DuMont,[34] and his pursuit of big pictures stimulated other manufacturers to similar tube-stretching efforts.

Network schedules were mainly evening affairs and many affiliates were not yet linked by landlines. These outlets were shipped their programming on a delayed basis via *kinescopes* – grainy film recordings taken off a studio monitor while the network show was airing live. When live or kinescoped network programming wasn't available, stations relied on old films from the 1930s (Hollywood refused to make more recent product available to its new competitor) and any local production they could think of to keep viewer attention and attract advertiser interest. Often these were little more than radio concepts and

personalities borrowed from a co-owned audio outlet. Even less exciting, "program directors abundantly scheduled test periods," Professor J.
Steven Smethers points out, "sometimes as filler material for occasional
holes in the program lineup or, more often (and oddly enough), as a
service to appliance dealers, who found the intricate designs of the test
pattern useful when adjusting the usually unstable picture of TV sets
manufactured in this era."[35]

Still, even by 1948, some major shows were mounted to more effectively captivate the public and build viewing habits. NBC's *Kraft Television Theatre* had debuted the previous year. This live program would
air more than 600 high quality dramas over the next decade. CBS
introduced *Toast of the* Town, a more broad-appeal variety program that
became *The Ed Sullivan Show* and ran until 1971. NBC countered with
Texaco Star Theater that introduced vaudeville and radio comedian
Milton Berle to the viewing audience. Berle's slapstick so dominated
Tuesday evenings that he was dubbed "Mr. Television" and almost sin-
glehandedly brought the medium to truly mass appeal. In 1949, Berle
would win one of the first Emmy awards given out by the new
Academy of Television Arts and Sciences as would ventriloquist Shirley
Dinsdale for *The Judy Splinters Show*, which bore the name of her
dummy. Judy spent a lot of time teaching child viewers how to brush
their teeth – no real coincidence given the prominence of toothpaste
advertisers on TV.

By this time, 40 percent of the US population was reachable by at
least one television station even though there were fewer than 2 million
sets in use. Viewership was much higher than the set penetration figure
suggests, because millions of people watched at the home of a neigh-
bor. Kids would crowd into that living room in the afternoon to view
Judy Splinters, Howdy Doody, Beanie and Cecil, and *Kukla, Fran, and Ollie*,
to be replaced in the evening by men watching boxing or couples
enjoying *Your Show of Shows* with Sid Caesar and Imogene Coca or
Mike Stokey's *Pantomime Quiz*. As in other countries, those first TVs
gathered people together in front of their flickering images like cave
dwellers must have been drawn to the first campfires. The phenomenon was new, it was exciting, and (once you completed the installment
payments on the set) it was free.

Though the total number of authorized stations was limited to 108
until 1952 when a four-year FCC "Freeze" on applications was
rescinded, the 1950s still started strong for television. Microwave and

coaxial relays now linked affiliates across the country allowing for further expansion of network programming. NBC's *Kate Smith Hour* became the first daytime program to be shown coast-to-coast and was soon joined by *The Garry Moore Show* on CBS. At night, a new crop of quiz shows such as *What's My Line?* and *Beat the Clock* on CBS shared the evening with migrant programs from radio like Groucho Marx's *You Bet Your Life* and *Your Hit Parade* on NBC and CBS's *The Jack Benny Show*. Lacking a radio operation from which to draw, DuMont had to develop original properties from scratch, the most notable of which was Jackie Gleason's comedy/variety extravaganza.

But the networks were no longer the only national distributors of programming. Frederic Ziv, who had begun a radio syndication business in the 1930s, brought his enterprise to television in 1950 with distribution of the half-hour Western, the *Cisco Kid*. Ziv's groundbreaking strategy of producing, acquiring and distributing series and telefilms direct to individual stations offered an alternative to network or locally-produced fare and established the mechanism by which independent (non-network-affiliated) outlets would build their schedules for decades to come.

The following year, the most successful television situation comedy of all time, *I Love Lucy*, made its CBS debut, sponsored by cigarette maker Phillip Morris. The show was filmed live before a studio audience using three cameras – a technique that subsequently became the standard for situation comedy production. *I Love Lucy* topped the ratings for several TV seasons and "went on to become the most widely syndicated comedy series in history."[36] Meanwhile on NBC, a much less heralded comedy program stretched the boundaries of the medium in other ways. Launched from Philadelphia, *The Ernie Kovacs Show* made ingenious use of visual devices, direct close-up conversations with the camera, and Dutch Masters cigar commercials that were as zany as the show they sponsored. Kovacs demonstrated that true television was not just radio in front of a camera, or a movie projected onto a cathode ray tube, but a unique technology. When he moved his show to CBS in late 1952, Kovacs pioneered the exploitation of special effects and became the first producer willing to spend thousands of dollars to achieve a four- or five-second shot.

Network programming now pushed into the early morning hours with NBC's *Today*. Hosted by the low-key Dave Garroway, who was frequently assisted by a rambunctious chimpanzee called J. Fred Muggs,

Today demonstrated that a day-opening blend of news, weather, interviews, and cast conversation could assemble a potent audience. As NBC executive Sylvester "Pat" Weaver, the program's creator, later observed: "At 7 A.M., you didn't want to rivet people to the set. You wanted them to tap in and out of the show, get dressed, brush their teeth, get ready to go to school or work, and know that whatever was new and important that day would be covered in the show."[37] Two years later, in 1954, Weaver brought network television to late night as well when he took what had been a local New York talk program starring Steve Allen and begat *The Tonight Show*. Through this vehicle, *Broadcasting & Cable* magazine recalls, the multi-talented Allen "inaugurated such talk-show staples as the opening monologue, the desk-chair-and-sofa set, bits that involved the studio audience, and bringing his cameras outside the studio for improvisation with the man on the street."[38] There is very little of today's talk show formula that is not a throwback to *The Tonight Show*'s first three years, or *The Steve Allen Show* that preceded it.

Meanwhile, doormat ABC was making innovations of its own after its merger with the United Paramount Theater chain that federal antitrust action had forced movie studio Paramount Pictures to divest. Paramount Theater executive Leonard Goldenson became the new ABC president and set about bringing film production to television. He first agreed to financially help Walt Disney build his Disneyland theme park in exchange for a 35 percent share of its ownership and promise of a weekly show from Disney – plus access to the Disney film library.[39] This cracked Hollywood's virtual boycott of the television industry. Soon Warner Brothers was producing for ABC as well, with its first success the hour-long Western series *Cheyenne*. The TV Western blossomed and Warner and other studios now set about creating a variety of film-for-television action and drama series. Before long, television production at some of these studios was outstripping their output of theatrical movies.

Television news was also coming into its own. Coverage of the 1954 Army/McCarthy hearings proved to be the downfall of Senator Joseph McCarthy who had manipulated public fears of Communism into a power-grabbing witch hunt. Edward R. Murrow's *See It Now* broadcasts had previously exposed the senator's smear and bullying tactics. The televised hearings confirmed Murrow's findings for all to see, thereby ending McCarthy's political career. Televised presidential news

Figure 3.4 CBS newsman Edward R. Murrow used television and his stellar reputation as a wartime radio correspondent to expose a senatorial demagogue and bring
world leaders into the American living room. Source: Photo courtesy of the National
Association of Broadcasters.

conferences began in 1955 when television actor and White House
adviser Robert Montgomery persuaded President Dwight D.
Eisenhower to begin the practice. These were not yet live events,
however. Each of Eisenhower's news conferences was filmed in the
morning so that Press Secretary James Hagerty could supervise editing
of the president's responses before their evening broadcast. Television's
international news, in contrast, did not easily lend itself to such
prepackaging. In 1957, when Soviet Premier Nikita Khrushchev was
featured in a taped Moscow interview on CBS's *Face the Nation*, and
Yugoslavian strongman Marshal Tito appeared on Edward R. Murrow's
See It Now, video's ability to create a global forum was clearly demonstrated (see Figure 3.4). At the end of the decade, television news also
proved willing to turn the cameras on its own medium as some of its
most popular game shows were revealed to have been rigged to build
ratings.

Quiz show riggings notwithstanding, TV's future looked bright as the medium consolidated and expanded its power at the expense of less nimble players. The network business became an industry of three as DuMont closed down in 1955; a victim of insufficient programming resources, loss of station affiliations, and a distribution system that was too heavily dependent on the troubled UHF band. Film exhibition suffered too, as the appeal of the tube seriously undercut the attraction of the silver screen. In 1957 alone, some 12,000 movie houses closed their doors. Of the 17,800 US theaters left in existence, one-third were drive-ins,[40] which offered young couples an environment for more participatory pleasures than just watching the movie.

Television at Center Stage

As the 1960s began, television's potency was again felt in the political arena. More candidates were spending significant sums of money on TV advertising, and exposure on news programs became essential to establish voter recognition. Senator John F. Kennedy proved the point in the four so-called Great Debates with Vice President Richard M. Nixon. Initially, Kennedy was a far less familiar figure to viewers than the Vice President, but their joint debate appearances brought him to recognition parity. In the first debate, on September 26, 1960, some 75 million people (the largest audience yet assembled) saw a relaxed and apparently tanned Kennedy in a dark suit flanked by a haggard, pasty Nixon clothed in washed-out gray who had refused to let CBS cosmetics artist Frances Arvold make up his face. Heard on radio, Nixon's responses seemed more knowledgeable and to-the-point. But on US TV screens, his drawn, almost devious, appearance spoke much louder than his words.

Some theorists attribute Nixon's extremely narrow election loss to this first debate and, thus, to his visual shortcomings. Whether Kennedy's victory margin was due to Great Debate #1 was less important than the fact that a number of politicians believed so. Kennedy became the first television president in the same way that Roosevelt had become America's first radio chief executive. In skillfully adapting to the new medium of his respective time, each man enhanced his own image as well as that of the communications vehicle that was maturing in time to carry that image. As Roosevelt had begun the live fireside chats, so Kennedy soon inaugurated the regular *live* TV press conference to establish a powerful and direct link between himself and the citizenry.

Scarcely 15 years since its postwar hatching, television was now a fully grown industry with almost 600 stations. In the early 1960s, virtually every US household had access to a set by which to view a wide array of film – and increasingly *taped* – entertainment programming. With the common availability of videotape, the risks inherent in live entertainment shows and commercials had largely been abandoned in favor of safer prerecording. On the other hand, newscasts attempted to avoid tape and "go live" as much as possible and were thus in place to cover the flights of astronauts and the aftermath of the Kennedy assassination. Audience measurement studies determined that more than 90 percent of US homes watched at least a portion of President Kennedy's telecast funeral and demonstrated that the video medium was now firmly positioned as a household necessity.

Only a few years later, television's new tools of color and satellite distribution were enlisted to cover the war in Vietnam. For the first time, viewers say battlefield carnage up close, in bloody living color, and with only a few hours' delay between the time the pictures were filmed and when they were aired following satellite relay to network headquarters. Some voices in the military complained that this enhanced communications system was undermining the national will to win. Other people welcomed it as a chance for the public at home to comprehend the real challenges faced by US armed forces in carrying out the administration's Southeast Asian policies. Both camps now recognized, as did television journalists, that video-encapsulated warfare had moved the psychological frontlines into the living room, setting off public reactions that could not always be predicted or controlled.

In early 1968, respected CBS newsman Walter Cronkite (who had taken the anchor job in 1963 just as network evening newscasts expanded from 15 minutes to a half hour) went on a fact-finding mission to Vietnam. This followed the Tet offensive in which the South Vietnamese capital itself had been infiltrated by the enemy. On February 13, Cronkite reported his unvarnished and pessimistic findings in a *CBS Special Report*. He concluded the broadcast with this statement:

It seems now more certain than ever that the bloody experience of
Vietnam is to end in a stalemate . . . On the off chance that military
and political analysts are right, in the next few months we must test the
enemy's intentions, in case this is indeed his last big gasp before negotiations. But it is increasingly clear to this reporter that the only rational
way out then will be to negotiate, not as victors, but as honorable people
who lived up to their pledge to defend democracy, and did as best they
could.[41]

Watching Cronkite's broadcast, President Lyndon Johnson turned to his
aides and said: "That's it. It's all over." One month later, Johnson
announced he would not be running for a second term. The power of
video journalism's pictures and most respected practitioner to change
the course of national policy had been amply evidenced.

The TV eye was casting a critical gaze on the domestic scene too.
For by 1968, the medium's technical capacity for graphically covering
real-life violence was being displayed in the cities of the United States
as well as Vietnam's rural hamlets. The assassination of civil rights leader
Martin Luther King and the riots that followed, the murder of Senator
Robert Kennedy after a hotel campaign rally, and the chaotic con-
frontations between Chicago police and protestors outside the
Democratic National Convention all made for television coverage that
was as compelling as it was anxiety-producing. Often, there was a tendency to blame the messenger for the event. As some military officials
were criticizing video for undermining the war effort, so Chicago
Mayor Richard Daley lambasted television's documenting of his police
pummeling protestors and news personnel alike. The following year,
Spiro Agnew, the new Republican Vice President, blasted the "small
band of network commentators and self-appointed analysts" who had
critiqued a Vietnam policy speech by President Richard Nixon and
asserted that "perhaps it is time that the networks were made more
responsive to the views of the nation and more responsible to the
people they serve."[42] Clearly, television was now taken seriously as a
national force. But this recognition was proving to be a mixed
blessing.

By the end of the 1960s, television was an integral part of the US
lifestyle and began to more fully service nonprofit objectives as well.
The Corporation for Public Broadcasting was now up and running to
assist the further development of noncommercial television. With the

help of the Ford Foundation and the Carnegie Corporation, the Children's Television Workshop took to the tube with *Sesame Street*, the first of its many projects to demonstrate that educational and self-esteem-building children's television could be as attractive and fast-paced as any commercial entertainment fare. Soon the Public Broadcasting Service (PBS) would be constituted with CPB assistance to serve as the autonomous interconnection for all national public television activities aimed at adults as well as children.

In fact, television was becoming such a core part of US and other cultures that few could recall how they ever did without it. To probe the depth of people's need for TV, German researchers at the Society for Rational Psychology in Munich paid a stipend to 184 habitual viewers for every day they abstained from watching. Despite the monetary incentive, one man went back to television after only three days, and no one held out for more than five months. While under video deprivation, subjects reported increases in moodiness, child spanking, and wife beating, with decreased interest and performance in marital sex. In exquisite seriousness, study director Henner Ertel postulated, "With people who watch regularly, many behavior patterns become so closely related to TV that they are negatively influenced if one takes the set away. The problem is that of addiction."[43]

Addiction, simple diversion, or something in between, television content by the mid-1970s was becoming ever more available and diversified. The three commercial broadcast networks still dominated US video. Yet independent stations were discovering that well-promoted local sports and movie offerings could lure audiences of a profitable size as counterprogramming to network half-hour and one-hour series. Such unduplicated offerings also made it more likely that cable systems would choose to carry an "indy's" signal beyond the limits of its over-the-air coverage pattern.

In the cable sector itself, Home Box Office demonstrated in 1975 that satellite carriage enabled cable networks to service systems everywhere in the country without the high cost and geographic limitations of land-based microwave distribution. The following year, independent broadcaster Ted Turner began using a satellite subsidiary to relay his Atlanta UHF station (WTBS) across the land and the *superstation* was born. Televangelist Pat Robertson followed in Turner's footsteps as his television station evolved into the Christian Broadcasting Cable Network – later ABC's Family Channel. Still, only about one-sixth of

US homes were as yet wired for cable by the end of the decade so broadcast television continued to reign supreme.

Then, during the 1980s, the competition floodgates opened wide. Beginning with Ted Turner's Cable Network News (originally debunked as the "Chicken Noodle Network"), the sports driven ESPN, and the unbridled Music Television (MTV), new and combative cable networks increasingly drained audience from broadcasters. Competition intensified even within the broadcast sector in 1986 when Australian media baron Rupert Murdoch bought six major-market television facilities from the Metromedia group. (Ironically, the core of this group consisted of former DuMont outlets.) He then completed his purchase of major program producer Twentieth Century Fox Film Corporation. With a key station group and a prominent studio/syndicator thus under his control, Murdoch launched the Fox network on October 9, 1986, when it beamed *The Late Show Starring Joan Rivers* to 99 affiliates and began the systematic expansion of its program schedule.

Still, the Big Three broadcast networks could look back on the 1980s with a great degree of satisfaction because the big programming hits all aired over their facilities. CBS's evening soap *Dallas* dominated the decade's first half; NBC's *The Cosby Show* comedy ruled the second. Big budget mini-series such as *Shogun*, *The Thorn Birds*, and *Lonesome Dove* (which all sprang from the explosive success of ABC's slavery-chronicling *Roots* in 1977) also drew huge audiences and ABC's *Monday Night Football* was the top sports draw. Creatively, Steven Bochco's *Hill Street Blue* pioneered edgy editing, sound, and multiple plotting techniques featuring flawed characters in frequently unresolved quandaries; elements that would reshape the fabric of much of television drama in the years to come.

Such success could not stem the tide of audience fragmentation, however. For one thing, CNN's 1991 domination of Persian Gulf War coverage demonstrated that *broadcast* network news was no longer supreme. On the entertainment front, cable networks like USA, Nickelodeon and Comedy Central were launched or expanded their viewership at broadcasters' expense. In 1994, Fox's infusion of $500 million into the 12 major-market New World Communications stations caused them to switch their affiliations from the "Big Three" to Fox and set off a ripple effect felt throughout the country. With the two new broadcast networks – the WB and UPN – that subsequently came on the scene in 1995, plus the Disney/ABC, Westinghouse/CBS, and Time Warner/Turner Broadcasting mergers, the video landscape now featured more choices then ever. These choices, however, were coming to be clustered under the corporate umbrellas of fewer and fewer players. A decade later, NBC had combined with Universal and the WB and UPN were merged into The CW network under the joint partnership of CBS Corp. and Time Warner.

Early in the new century, hundreds of different "channels" were available to and through cable. Begun in the 1990s, direct-to-home satellite service (DBS) was now a robust competitor to cable offering a similarly wide array of program services. These channels' content ranged from the kid-friendly fare on The Disney Channel and Nickelodeon to the high-quality, adult drama produced for HBO and the low quality but profitable soft-porn offered up by other pay services. Increasingly, the same show could be made available continuously through the replay capabilities of VOD (video-on-demand) features that reside in a set-top box or on a program-provider's server. A show now could compete with other viewing options not only as it is first aired, but continuously. Clearly, the medium had come a long way from the tidy three-network universe that had built US television.

Content for the Second and Third Screens

In the twenty-first century, television programming has more expansively become *video* programming and is no longer exhibited only on a television receiver. Broadband technology means that programming can now be delivered over the Web to the "second screen" – the computer. Sometimes this is simply an alternate avenue for conventional programming being aired simultaneously over television (and radio) stations that are "streaming" their signals via the Internet. But programming interests are also packaging content unique to the Web that may be ancillary to existing television offerings or separate program properties entirely. Such shows take advantage of the Internet's flexibility. They can be constructed in unconventional lengths and feature multiple endings. They can tap into content made popular by video games. And they can be archived and shared from one computer to another, subject only to each box's memory limits, the speed of the broadband pipe to which it is connected, and the frantic efforts of copyright holders to protect the content they have created.

Programming on the cell phone's "third screen" opens up further possibilities. In addition to news and weather snippets, new entertainment vehicles like "Mobisodes" are borrowing a format pioneered by the ten-minute radio serials of the 1930s and updating it for the visual entertainment of a cell phone-toting populace. The technology is decidedly different than when radio first lured audiences. But the demand by people for instant electronic information and entertainment programming relevant to their lifestyle and tastes has remained a constant that must be served. As Edwin Artzt, CEO of giant packaged-goods manufacturer Proctor and Gamble astutely observed back in 1995:

> Content will always be king. It's easy to get distracted by new-media technologies, but once the novelty of interactive entertainment or pay-per-view services wears off – and it will wear off quickly as these technologies become pervasive – what will count with consumers is what the programming is.[44]

Chapter Rewind

Radio was transformed by wireless experimentation into a self-standing mass communications vehicle in the 1920s. The public's imagination was captured by the entertainment and information it could pick off the air with preassembled receivers now available in stores. WEAF's sale of airtime demonstrated that radio could be a profitable consumer business and large numbers of stations were soon launched. Beginning with AT&T, networking proved to be an efficient way to provide national programming to local audiences. By 1929, radio was already a healthy $40 million business with three networks and the first hit show, *Amos 'n' Andy*. Even the onset of the Depression did not stop the medium's growth because people clung to it as the only form of entertainment they could afford. Newspaper publishers now perceived radio as a competitor for advertising and restricted the amount of wire service material made available for broadcast. Radio countered with its own news cooperatives and was bolstered by President Franklin D. Roosevelt's frequent use of the airwaves for direct communication with the citizenry.

Programming in the 1930s ran the gamut from outstanding cultural presentations to lowbrow, disparaged fluff but radio's immediacy made it indispensable in monitoring the gathering war clouds. Once the United States entered World War II, radio marshaled its resources to help the war effort through news coverage and morale-building entertainment that reflected war-related themes and concerns. After the war, radio encountered a host of problems. FM suffered from a lack of interest on the part of advertisers, listeners and even broadcasters themselves. As the network companies and other major operators shifted their resources to the new television medium, radio became less national and more local with deejays replacing network shows and stars.

From the mid-1960s, radio formats became more fragmented. FM grew to equal and then surpass AM with younger listeners avoiding "antique modulation" entirely. Once satellite technology became available, consolidating syndicators and networks offered stations everything from brief features to entire formats while the 1996 Telecommunication Act fostered the creation of huge radio groups. In the new century, direct-to-consumer satellite radio poses new competition for local stations.

Though television was briefly available to the public beginning in 1939, the war postponed significant development until 1946. Early shows were often primitive, but the novelty of the medium meant that the first neighborhood household to obtain a TV set became the gathering place for the entire block. Until interconnections were constructed, networking was accomplished by the shipping of filmed *kinescopes* of live programs. Network newscasts were only 15 minutes in length, but television's journalistic power was demonstrated in the exposure of Senator Joseph McCarthy's inquisitorial brutality. By the end of the 1950s, television was a fixture in 75 percent of US households and Hollywood began to consider the medium more as customer than competitor.

Politicians like John F. Kennedy learned how to effectively utilize the new medium while the coverage of civil rights demonstrations and the Vietnam War showed that television had a political power in its own right. The rise of independent stations and cable began to fragment the TV audience by the late 1970s. Program choice – both free and pay – widened. But industry consolidation meant more of these choices were controlled by fewer corporate players whose shows were carried to consumers by stations, cable, and DBS. In addition, conventional and new forms of programming began to lure audiences from the TV set to the second and third screens of the computer and the cell phone.

SELF-INTERROGATION

1   What factors converted radio from a "wireless telephone" to a mass consumer medium?
2   Compare and contrast how NBC and CBS were formed.
3   What made FDR the first "radio president" and JFK the first "television president"?
4   Why and through what mechanism did the newspaper industry restrict the flow of news copy to radio?
5   Why did William Paley of CBS lure radio personalities away from rival networks in the late 1940s? What financial mechanism did he use to accomplish this?
6   What was the Progressive format and what changes to the radio industry did it help to stimulate?
7   What were the factors that inhibited the growth of US commercial television from 1939–52?
8   What was DuMont and what role did it play in the growth of television?
9   How did Hollywood respond to the coming of television? How did the construction of Disneyland help to change the movie industry's antagonism toward the new medium?
10 How and why did the television industry change after HBO pioneered satellite distribution of its programming?
11 What are the second and third screens and how might they impact video programming?

NOTES

1   Frederick Lewis Allen, *Only Yesterday* (New York: Bantam Books, 1959), 55.
2   "College Lectures by Radio," *Literary Digest* (May 13, 1922), 28.
3   "Radio's Increasing Value to Public," *Editor and Publisher* (November 20, 1922), 143.

4 "Radio Cutting down Church Attendance by Broadcasting Services, Says Bishop," *New York Times* (May 27, 1923), II/1.

5 "Review of the Year 1923," *Scientific American* (January 1924), 14.

6 David Sarnoff, "Two Schools of Thought," *Nation* (August 1924), 12.

7 Iris Selinger, "Master Builder of Brands," *Advertising Age* (April 4, 1994), A-3.

8 O. E. Dunlap, "Shall Advertising Be Given the Air?," *Outlook* (November 11, 1925), 387–88.

9 "Radio Conference Ends," *New York Times* (December 20, 1925), 8/13.

10 John McDonough, "Radio: a 75-Year Roller-Coaster Ride," *Advertising Age* (September 4, 1995), 23.

11 Robert Landry, *This Fascinating Radio Business* (Indianapolis: Bobbs-Merrill, 1946), 63.

12 Reed Bunzel, "Filling the Magic Box," *Broadcasting* (December 9, 1991), 27–8.

13 Harrison Summers (ed.), *A Thirty-Year History of Programs Carried on National Radio Networks in the United States, 1926–1956* (Columbus, OH: Ohio State University Press, 1958), 21.

14 George Manning, "Average Radio Cost Is $310 per Hour," *Editor and Publisher* (November 15, 1930), 13.

15 Robert Mann, "Radio 'Spotlight' Copy Is Increasing," *Editor and Publisher* (June 14, 1930), 13.

16 "Genesis of Radio News: The Press-Radio War," *Broadcasting* (January 5, 1976), 95.

17 John McDonough, "TV's Seismic Shift," *Advertising Age 50 Years of TV Advertising* (Spring 1995), 4.

18 "Yardstick to Radio," *Time* (September 9, 1935), 41.

19 Robert Metz, *CBS: Reflections in a Bloodshot Eye* (New York: Signet, 1976), 4.

20 "Blurbs in the Breaks," *Business Week* (January 16, 1937), 35.

21 Donald O'Toole, "Remarks in the House of Representatives," *Congressional Record* (December 1937), 23891.

22 Llewellyn White, *The American Radio* (Chicago: University of Chicago Press, 1947), 75.

23 "1941," *Broadcasting* (December 22, 1980), 101.

24 "Outspoken Broadcast," *Time* (August 9, 1943), 62.

25 William Robson's comments on CBS commemorative recording of *Open Letter on Race Hatred*.

26 Bunzel, "Filling the Magic Box," 29.

27 McDonough, "Roller-Coaster Ride," 24.

28 Michael Keith, *Radio Programming: Consultancy and Formatics* (Newton, MA: Focal Press, 1987), 89.

29   Dick Osgood, *WYXIE Wonderland* (Bowling Green, OH: Bowling Green University Popular Press, 1981), 485.

30   Robert Mullins, "XM? Sirius? Radio is Fighting Back," *Silicon Valley/San Jose Business Journal online*, June 24, 2005.

31   Jane Dalzell, "Who's On First," *Advertising Age 50 Years of TV Advertising* (Spring 1995), 8.

32   "60 Years of Programming," *Broadcasting 60th Anniversary Issue* (December 9, 1991), 30.

33   "What Happened to the Dream World?," *Fortune* (February 1947), 93.

34   *The Story of Television*, promotional film issued by DuMont Television, ca. 1950.

35   J. Steven Smethers, "Unplugged: The Growth of Midwestern Television Before Network Interconnection, 1949–1952," paper presented to the Association for Education in Journalism and Mass Communication 1995 Convention, Washington, DC, 13.

36   Michael Logan, "Ay-yi-yi! The Night America Fell in Love with Lucy and Desi," *TV Guide* (October 12, 1991), 13–14.

37   Neil Hickey, "Today's Yesterdays: 40 Years of Chats, Spats and Changing Hats," *TV Guide* (January 11, 1992), 14.

38   Dan Trigoboff and Beatrice Williams-Rude, "A Class Act On-Screen and Off," *Broadcasting & Cable* (November 6, 2000), 19.

39   Steve McClellan, "ABC's Goldenson Dead at 94," *Broadcasting & Cable* (January 3, 2000), 14.

40   *Americana Annual, 1958* (New York: Americana Corporation, 1958), 505.

41   Peter Braestrup, *Big Story: How the American Press and Television Reported and Interpreted the Crisis of Tet 1968 in Vietnam and Washington* (Garden City, NY: Anchor Books, 1978), 134.

42   Excerpts from the released text of Vice President Spiro T. Agnew's speech to the Midwest Regional Republican Committee (Des Moines, IA), November 13, 1969.

43   "Life without the Tube," *Time* (April 10, 1972), 47.

44   Edwin Artzt, "Artzt Enthusiastic About CASIE Gains," *Advertising Age* (March 13, 1995), S-24.

CHAPTER 4

Regulatory Chronicles

Louis A. Day

Advances in communications technology, from the invention of writing by the ancient Sumerians more than five thousand years ago to the construction of the Internet, have almost always been accompanied by some government regulation. The focus of this book, of course, is electronic media. This chapter contributes to this focus through an exposition of the US regulatory scheme that attempts to strike a middle ground between government control and a policy of *laissez-faire*. Space does not permit an exhaustive treatment of this topic, but you should emerge from this discussion with a deeper appreciation for the philosophy and legal principles that animate government regulation of the electronic media.

Early Electronic Media Regulation

The invention of the telegraph in the nineteenth century heralded the inauguration of the world's first medium of electronic communication. In contrast to their American counterpart, the telegraph services in all the European states, except for England, were government controlled. European political leaders quickly recognized the telegraph's promise as a cross-border medium of communication, and in 1865, 20 countries convened in Paris to draw up an agreement on telegraph rates, secrecy of communications, and the establishment of an international organization, the Telegraph Union, to handle the administrative details of these matters. But even as the international community embraced the novelty of the electric telegraph, wireless communication emerged as a distinct alternative.

As mentioned in Chapter 2, the first radio patent was issued to Guglielmo Marconi in 1896, and through the formation of wireless companies in Great Britain and subsidiaries in other countries, Marconi gained a monopoly on most patents for radio equipment. He tightened his monopoly by supplying certain vessels with wireless equipment on the condition they not communicate with vessels employing other types of apparatus.[1] The difficulties this posed for international shipping were apparent. Therefore, the German government convened a series of conferences that ultimately resulted in the drafting of a convention and the establishment of the Radiotelegraph Union that would, among other tasks, assign frequencies to avoid interference, organize radio traffic, and promulgate technical standards for equipment and operators. In 1932, the Radiotelegraph Union was replaced by the International Telecommunication Union (ITU), which united the technical oversight of all forms of electronic media – telegraph, telephone, and radio – under one regulatory umbrella organization.

Radio Regulation's Foundation in the United States

The unregulated environment of radio communication between 1906 and 1912 would have been familiar to today's Internet generation as hundreds of amateurs built radio sets and sent messages to each other, some of which were obscene. It was a stray iceberg in the North Atlantic that finally brought some degree of civility and sobriety to this fledgling radio industry. On the night of April 12, 1912, the ocean liner *Titanic* sank as the radio traffic that swirled about the disaster was characterized by "ceaseless interference, cruel rumors, and misleading messages that filled the air from unknown sources during the disaster."[2] Although a federal law enacted in 1910 required all passenger ships to carry a wireless set,[3] the electronic chaos undermined the *Titanic's* ability to contact would-be rescuers. In the aftermath of this human tragedy, the press denounced radio amateurs, and the public

demanded quick government action to bring some order to radio communication.[4] The result was the Radio Act of 1912, which required all
users of the radio spectrum to obtain a license and placed licensing
authority in the hands of the Secretary of Commerce. In addition, the
statute allocated the spectrum according to classes of usage (e.g., amateurs, military, and commercial) and assigned priorities for access to the
airwaves, with emergency signals being given top preference.[5]
However, the Secretary of Commerce had no discretionary authority
in issuing licenses, which set the stage for further chaos.

Until 1917, most of the radio activity in the United States was conducted by amateurs who amused themselves by transmitting a blend of
recorded music and commentary to other amateurs who possessed
compatible receiving equipment. The government tolerated such
experimentation as long as it did not interfere with the more publicly
significant functions of radio communication. But this *laissez-faire* attitude came to an abrupt halt with the entry of the United States into
World War I. Suddenly, radio was viewed as a vital component of the
war effort, and the Navy was given extraordinary powers to harness
radio's potential by requiring that manufacturers pool their patents and
cooperate in future research efforts.[6] The use of wireless communication as a military tool had an unintended impact on radio policy.
Because many American servicemen were trained in wireless, they
returned to civilian life with an impressive grasp of and enthusiasm for
wireless broadcast.[7]

The first commercially licensed broadcast station, Pittsburgh's
KDKA, began formal operations on November 2, 1920, with a broadcast of the Harding-Cox Presidential election returns. This political
coverage did not precipitate a rush on license applications, but the
broadcast the following year of the World Series between the Yankees
and the Giants did.[8] By December 1922, there were more than 21,000
radio transmitting stations of all kinds on the air.[9] There were millions
of listeners and far too many stations. Like the highway system, the
electromagnetic spectrum is a finite commodity, and in the 1920's
the radio industry was close to electronic gridlock.

The administrative impotence of the Secretary of Commerce to
regulate the resulting electronic chaos was made manifest when a
federal court ruled that a station owner could not be punished for
ignoring a frequency assignment made by the Secretary,[10] and the
Attorney General declared that the Secretary had no authority to limit

frequency, power or time used by any station.[11] Congress initially ignored pleas from the Secretary to provide him with broader authority to regulate the mounting interference. In Congress' defense, the nature of broadcasting was still undefined, and "it was difficult to pass a law to regulate an unknown quantity."[12] But by 1926 the situation had become so untenable that broadcasters themselves demanded government intervention. Their entreaties were coupled with pressure from leaders in government, education, religion, industry and labor, prompting Congress to enact the Radio Act of 1927, the first law to extend federal oversight to the fledgling broadcast industry. The 1927 Act constituted broadcast regulation's foundation for the next several decades.

Like the Radio Act of 1912, the new statute stipulated that a license was required to operate on the electromagnetic spectrum. However, it removed the licensing function from the Secretary of Commerce and vested it in a new five-person Federal Radio Commission (FRC), whose terms were extended on a year-by-year basis. Congress rejected private ownership of the airwaves and declared the spectrum to be government property. Nothing in the nature of the medium or the electromagnetic spectrum made this choice inevitable, but there was little consideration of competing models, as evidenced in this representative comment from Senator Clarence Dill: "Under the Constitution, the Government has power to regulate interstate commerce" (see Figure 4.1). And because radio is inherently interstate rather than intrastate, "[t]he one principle regarding radio that must always be adhered to, as basic and fundamental, is that government must always retain complete and absolute control of the right to use the air."[13]

In exchange for a license, broadcasters were entrusted to serve the "public interest, convenience, or necessity," a phrase that was imported from the language of public utilities and transportation regulation. Ironically, the adoption of the public interest standard as a matter of public policy served to marginalize minority viewpoints and programs that the FRC did not consider socially responsible. For example, the FRC attacked "propaganda" stations, warned a New York socialist station that it must "operate with due regard for the opinions of others."[14] It revoked the licenses of a Los Angeles station because of "sensational" attacks on public officials and Roman Catholics[15] and a Kansas City station because of efforts by medical quack John R. Brinkley to promote his bizarre theories of revitalizing middle-age male sexuality.[16] These and similar actions were generally upheld in the federal courts,

Figure 4.1 Senator Clarence Dill of Montana, co-sponsor of the 1927 Radio Act and chief advocate of an independent commission to regulate broadcasting. Source: Courtesy of The Library of Congress.

but the FRC's position "may have been strengthened by the fact that the broadcasters involved were rather notorious characters with long and well-documented histories of using their stations for their own personal benefit."[17] Nevertheless, the ambiguity of the *public interest standard* produced confusion and uncertainty within the broadcast industry, a condition that continues to the present.

Because the 1927 Act focused both on licensing and content, there were clearly First Amendment implications for the government's regulatory scheme, but this issue received scant public attention. There were at least three reasons for subdued interest in radio as a constitutionally protected medium. First, although educators, labor and religious leaders saw radio as a noncommercial medium with great potential for the dissemination of ideas and information, their overtures were drowned out by a drumbeat of more powerful voices from the fledging broadcast industry that campaigned aggressively for a more orderly spectrum, thereby permitting them to proceed with the commercial exploitation of the radio enterprise. Second, the Supreme Court's free speech jurisprudence was still in its primitive stages. While some cases had been decided in the aftermath of World War I, constitutional scholars and even the Court itself were struggling to comprehend the dimensions of this constitutional guarantee. Third, most broadcasters and listeners perceived of radio as mainly a commercially-based entertainment medium. The existing primary entertainment media of movies and the circus did not enjoy First Amendment protection, and radio appeared to be more analogous to them than to the traditional print media.[18]

The 1927 Act was superseded seven years later by the *Communications Act* of 1934, which established the Federal Communications Commission to replace the FRC. While the FRC had been created as a temporary body to manage the spectrum, the FCC was given broad, permanent authority over all forms of telecommunication, not just radio. The FCC consisted of seven commissioners (the current number is five), appointed by the President, with a limit of four from any one political party. The body was given broad statutory authority to issue broadcast licenses (the original term was three years) and to ensure that licensees operated in the public interest. Failure to conform to this standard could result in fines, short-term license renewals, and in the most egregious cases, license revocation.

The FCC was the ultimate manifestation of the regulatory state, which reached its zenith during President Roosevelt's New Deal. The *laissez-faire* economic ideology of the Nineteenth Century gradually gave way to theories that advocated some government control over the private market. Through this view, private enterprise was left to its own devices except where government regulation was justified in those industries *affected with a public interest*. Broadcasting was such an industry.

But this duality of private ownership of an industry "affected with a public interest" and subject to government regulation produced inevitable constitutional objections from licensees. Congress paid homage to First Amendment concerns through Section 326 of the Communications Act prohibiting FCC censorship, that is, *no prior restraint*, of program content. Yet at the same time the Commission was granted broad discretionary authority to review the quality of licensee performance, which raised the specter of administrative sanctions for failure to live up to the rather vague public interest standard. For broadcasters, the prohibition on pre-censorship – but the possibility of subsequent punishment for failure to comply with government mandates – was a distinction without a difference. Nevertheless, this disparate constitutional treatment of broadcasters is consistent with the Supreme Court's jurisprudence, which holds that the distinct characteristics of different media justify different First Amendment approaches to regulation.[19]

Initially, the scope of the FCC's authority in regulating the broadcasting industry was not entirely clear. Broadcasters viewed the Commission as a traffic cop whose primary function was to allocate frequencies to avoid interference and technological chaos. Any regulation that targeted content regulation, in their judgment, would violate the free speech clause of the Constitution. The FCC, on the other hand, viewed its mandate more expansively, arguing that the public interest standard required some oversight of a licensee's programming policies and performance.

Thus, based upon its public interest mandate, the Commission in 1941 promulgated its *chain broadcasting regulations* that were designed to diminish the radio networks' (CBS and NBC) dominance over their affiliates and to restore some measure of autonomy to local licensees in program decision-making. The Commission proposed to deny a license to any applicant that did not conform to its regulations. Seeking to preserve their programming power, the networks challenged the agency's authority under the Communications Act to promulgate such regulations and its constitutional authority to deny a license for a station's failure to operate in the public interest. Two years later, in a case that affirmed the Commission's statutory and constitutional authority to regulate the contractual relationships between the radio networks and their affiliates, the Supreme Court[20] ruled that the denial of a license for failure to operate in the public interest is not a violation of the First Amendment. According to the Court, the plain language of the Act placed an affirmative obligation upon the FCC to ensure an effective (i.e., qualitative) use of the scarce broadcast spectrum. To this day, the frequency scarcity rationale remains the only legal justification for the regulation of programming in the United States, although other rationales have been suggested periodically.[21]

Administrative Structure for Broadcast Regulation

The *Federal Communications Commission* is one of the largest of the more than 50 independent agencies that reside within the federal bureaucracy. These agencies blur the line between the separation of powers in such a way that would be unconstitutional if performed separately by any of the three branches of government. Their rule-making authority invests them with a legislative function. Their enforcement powers provide them with an executive prerogative. And their adjudication of disputes and rules violations, often in public hearings presided over by an administrative law judge, bears a striking resemblance to the judiciary function.

The FCC currently consists of five commissioners, appointed by the President for five-year terms with the concurrence of the Senate. The President appoints one of these commissioners to be chair. The Commission is supported by a large staff allocated among six bureaus, one of which is the Media Bureau which has oversight responsibility for broadcasting and cable, and ten staff offices.

The Communications Act provides the FCC with a long list of responsibilities, ranging from supervision of the technical aspects of the broadcast industry to certain kinds of content regulation. However, at the risk of oversimplification, the Commission's functions as they pertain to broadcasting can be broken down into three areas: licensing, rule-making, and adjudication of licensing disputes and rules violations.

Licensing

The nucleus of the Commission's power is in the *licensing function*. No one can legally operate a broadcast station without a license, and this basic reality constitutionally differentiates the radio and television media from other forms of mass communication. The original license period was three years, but in the Telecommunications Act of 1996, Congress extended the term to eight years.

According to the Communications Act of 1934, the FCC may issue a license only if it is convinced that the applicant will serve the public interest; a highly subjective and ambiguous standard. For the first four decades of the act's enforcement the documentation required to support an affirmative finding in this respect was fairly extensive. Until the 1980s, for example, the Commission required license applicants and licensees to conduct elaborate ascertainment surveys to determine community needs and to advance a plan for how these needs would be met. However, in the deregulatory environment of the 1980's the Commission dropped such formal requirements. Ascertainment studies are still sometimes used to convince the FCC that a license proposal is in the public interest.

While most license applications are fairly routine, it is not unusual to have competing applications for the same frequency. Before the FCC simplified the process, comparative hearings were frequently held to weigh the relative merits of the various applicants. This was often a tedious and lengthy undertaking, involving the filing of a myriad of documents and written testimony and the weighing of evidence by an administrative law judge. But since 1999, pursuant to Congressional authorization, the Commission has used a lottery to award licenses when there are competing applicants. There is also a fairly sophisticated electronic auctioning system to handle the allocation of a number of related licenses. And the licensing process itself has been greatly simplified in terms of the amount of paperwork that must be filed.

Once a license is issued, license revocation or non-renewal, which can result in the loss of millions of dollars in investment, is still relatively rare. Instead, the more common sanctions for rules violations and failure to operate in the public interest are fines or short-term license renewals. Because of constitutional problems, program content itself is seldom the sole justification for non-renewal. But objectionable programming can cost a station its license when coupled with a lack of management supervision or an attempt to conceal information from the Commission.[22]

Rule-making

While the Communications Act of 1934, as amended, specifically addresses certain policy matters pertaining to the regulation of broadcasting and cable (such as the provision relating to political broadcasts, discussed later), it is FCC rules and regulations that constitute the bulk of the government's regulatory apparatus pertaining to the electronic media. The rule-making procedure is fairly complex and must provide for public comment before any proposed regulation is formalized. The Act gives the Commission broad statutory authority in implementing the public interest standard, but the agency may not abuse its discretion in doing so. In other words, there are limits. The federal courts frequently defer to the FCC's technical expertise on a disputed matter, but they still insist that a Commission rule be based upon a reasonable application of one or more provisions of the Communications Act. In other words, FCC rules must not be *arbitrary* or *capricious*.

The adjudication function

Like other administrative agencies, the FCC must adjudicate licensing disputes and violations of the Communications Act and its rules and regulation. In so doing, it has an arsenal of *sanctions* at its disposal, including fines, short-term license renewals, and in extreme cases, license revocation. When a complaint against a station is received, the FCC will request the licensee to respond in writing. The mere fact of the inquiry itself may be sufficient to bring the station into compliance. This is sometimes euphemistically referred to as "regulation by raised eyebrow." If the Commission is satisfied with the licensee's response, the proceeding may be terminated at that point. If not, it then serves the licensee with a Notice of Violation. The licensee may elect to acquiesce in the Commission's findings and accept some form of sanction, which may include an apology to the offended party, a letter of reprimand, a fine, or a cease and desist order. If the offense is one for which a fine may be imposed, the FCC also issues a Notice of Apparent Liability to Monetary Forfeiture, to which the licensee has ten days to respond with a written answer.[23] If the station contests the Commission's conclusions, a formal hearing is held before an administrative law judge (ALJ), who may reject or affirm the agency's findings. At every stage of the proceedings, the licensee is entitled to due process. The licensee may appeal an ALJ's adverse ruling to the full five person commission and beyond that to the federal court of appeals (usually the Washington, DC, circuit) and perhaps even to the US Supreme Court.

Television and Technological Standards

The Communications Act specifically authorizes the FCC to promulgate technical regulations to ensure the efficient operation of the telecommunications industries. However, on a more cosmic scale one of the agency's indispensable roles is to *coordinate* the introduction of new technologies. The Commission's decisions have not always resulted in a seamless transition from old technologies to new, as exemplified by the emergence and evolution of television. The relative simplicity of radio technology, for example, allowed for its natural evolution and development as a medium of mass communication. But television's complicated transmitter-receiver relationship necessitated a more calculated approach to full governmental approval of the rapidly emerging video medium. In particular, this involved selection of a standardized "line system," the process by which the electronic scanning of images by the television camera is synchronized with TV receivers.[24] As Chapter 2 discussed, RCA was one of the early manufacturers of TV sets, and in 1940 an RCA-led group tried to convince the FCC to adopt the 441-line electronic system that it had unveiled at the 1939 New York World's Fair. However, the Commission was skeptical of this system's technical quality and thus established an industry-wide committee of engineers, the National Television System Committee (NTSC), which ultimately rejected the 441-line system in favor of a higher quality 525-line system that for several decades has remained the US standard.

In the 1980s the FCC began considering the role that the emerging digital technology might play in the American broadcasting system. This digitally-based system would eventually become known as *high definition television* (HDTV). By 1997, the FCC had enough confidence in this new technology to order a gradual phase-in of HDTV under a plan that would allow stations to continue to broadcast analog signals while acquiring a new digital channel. The Commission set a number of deadlines for activation of these digital channels and abandonment of analog ones but these mandates proved to be unrealistic. The year 2009 was eventually established as a more attainable deadline for the switch. Further, to maximize the probability of digital penetration, the FCC gave set manufacturers until 2007 to include digital tuners in all TV sets.

In television's early days, another problem confronting the Commission was the allocation of frequencies and a plan to provide for an equitable distribution of TV service. Although television technology was developed in the 1930s, World War II put a temporary halt to the cultivation of this medium. Following the war, the number of applications for stations increased dramatically. The FCC had approved channels 2–13 (VHF) for the new service, but it did not have a plan in place to accommodate the demand for licenses. The Commission therefore imposed a freeze on television station authorizations to avoid a repeat of the chaos that had plagued radio broadcasting in its formative stages. The freeze, which was originally predicted to last six months, lasted for four years.

During the four-year freeze the FCC devised an allocation table that assigned stations by geographic areas and also opened seventy channels in the *Ultra High Frequency* (UHF) band, the technical characteristics of which were still unclear. The FCC had hoped that UHF outlets would soon achieve parity with the VHF stations, but the weaker signal and the lack of sets equipped to receive UHF quickly relegated UHF stations to second-class status. They were further victimized by a vicious cycle. Viewers could receive UHF stations only by investing in a converter, but they were unwilling to do this unless they were convinced that UHF stations could offer interesting programs. Unfortunately, such programs are expensive and advertisers were unwilling to support such fare unless they were guaranteed a sizable audience. In addition, the established VHF stations had already locked up network affiliations and hence the best shows. Their UHF competitors had to depend upon syndicated material and local talent.[25]

To overcome viewer resistance to the purchase of UHF converters, in 1962 Congress amended the Communications Act. It authorized the FCC to require both a UHF and a VHF tuner on all TV sets. Though UHF stations would continue to struggle for a number of years, this legislation helped them eventually emerge from competitive obscurity.

While the FCC was struggling with the allocation and frequency problems of the new medium, it was forced to consider standards for a related innovation: color television. In 1950, the choices were a CBS-backed system or an RCA system. The FCC initially selected the CBS system, believing it to be somewhat superior to its competitor, but CBS's protocol was not compatible with existing monochrome standards. The industry looked askance at a color system that might soon be discarded.[26]

Two events provided some much-needed breathing room for the Commission to reflect upon its approval of color TV standards, First, RCA filed suit against adoption of the CBS system, which delayed matters as the case wound its way to the Supreme Court. The Court eventually sided with the Commission.[27] Second, the government asked manufacturers not to make color TV sets during the Korean War emergency. During this delay the National Television System Committee (NTSC) went to work on the problem. In 1953, based upon NTSC recommendations, the FCC adopted new *compatible* color television standards (through which the same picture could be received on both color and black and white sets). This system had been perfected by RCA during the time it bought by tying up the CBS system in court.

The Structure of the Industry

As the power and influence of the radio industry began to emerge in the 1930s, the FCC took an activist posture in preventing unreasonable concentrations of station ownership and network dominance. The Commission's policy oversight of the structural aspects of the electronic media has been guided by the separate but related goals of promoting competition and diversity. These goals flow directly from First Amendment principles and the public interest mandate set forth in the Communications Act of 1934. In the Commission's view, too many stations in too few hands is contrary to the spirit of ownership and programming diversity envisioned by the Communications Act. Nevertheless, this philosophy has been sorely tested beginning in the 1980s when Congressional and industry pressure convinced the FCC to modify the ownership caps. This demand to alter the ownership rules occurred coincidentally with a federal judiciary that was less inclined to defer to the Commission in its decision-making role. The courts began requiring the Commission to explain specifically how their ownership rules served the public interest. Relying upon the "diversity of broadcast voices" argument was no longer sufficient.[28]

The FCC's *ownership rules* are essentially of three kinds: (1) those that limit the number of broadcast outlets that one individual or corporation may own within a given market or service area; (2) those that place limits on national ownership coverage; (3) those that restrict cross-media ownership (such as controlling both a newspaper and broadcast station in the same market). The original rules limiting the nationwide number of licenses under common ownership in the various broadcast services were promulgated in the early 1940s. By 1954, the Commission had set the per owner limit at 7 AM, 7 FM, and 7 TV; a cap that remained unchanged for three decades. In addition, they prohibited any licensee from owning or controlling more than one station in each category in a given market. In 1975, the Commission imposed restrictions on the cross ownership of newspapers and broadcast properties within the same market, although most existing combinations were grandfathered (that is, the licensees were not required to divest themselves of either their newspaper or broadcast properties).

It is risky to describe specifically the current ownership rules because they have undergone several changes in the past two decades and have increasingly been caught in the legal crossfire between the supporters of ownership liberalization and critics who view such policy changes as a threat to media diversity. In essence, however, the Commission's rules adopted in 2003: (1) limit the number of TV stations a single individual or company may own based upon the percentage of the total national viewing audience served by those stations; (2) abolish the cap on radio station ownership; (3) restrict the ownership of both radio and TV stations in a single market, based upon the number of stations in that market; (4) regulate the co-ownership of television and radio stations and newspapers in the same market based upon the number of media properties located there.

The new rules prompted an unprecedented deluge of public criticism, according to FCC Commissioner Jonathan Adelstein, as 750,000 people wrote, called, faxed, and e-mailed the FCC. An overwhelming 99.9 percent of these communications were opposed to further media consolidation.[29] In June 2004, a federal appeals court remanded the

rules back to the FCC, declaring that the Commission had not provided sufficient justification for its new ownership limits.[30]

Content Regulation

Despite the Communication Act's admonition against program censorship (Section 326), the statute's command to broadcast licensees to
operate in the public interest has occasioned the FCC's periodic
appraisal of the industry's programming performance. The Commission's first serious attempt at program regulation occurred in 1946 with
the issuance of its *Blue Book*, so-called because of the color of its
binding. Officially entitled *The Report on Public Service Responsibility of
Broadcast Licensees*, this publication set out standards that helped to shape
program regulation for several decades. Although the measures in the
Blue Book were never fully enforced, it rebuked broadcasters for their
lack of local programs, overabundance of commercials, and overall
failure to serve the public interest. The *Blue Book* exhorted licensees
to remedy these deficiencies through the origination of more local programs and the inclusion in their broadcast schedules of sustaining
programs (those without commercial sponsorship).

Foremost among the Commission's concerns was the need for
broadcasters to serve their own communities; a reasonable expectation
that flowed from the idea of *localism* fundamental to the American
broadcasting system. However, the fulfillment of this goal presupposed
licensee familiarity with the needs and problems of their communities
and the development of programming to serve those needs. This expectation was formalized for commercial stations in 1971[31] and four years
later for non-commercial stations when the FCC announced that
licensees would be required to ascertain community needs. Specifically,
ascertainment consisted of surveying the general public and community
leaders, analysing the results and then identifying the ten major problems confronting the community. Finally, the station must propose and
broadcast programs that would address those needs. The ascertainment
process soon became so burdensome on broadcasters in terms of time
and paperwork that in 1981 the Commission began scaling back the
requirements – and subsequently eliminated formal ascertainment procedures altogether. Nevertheless, stations are still required to compile a

list of five to ten issues of importance to the local community and to maintain that list in their public file.

Broadcast licensees are bound by a myriad of programming responsibilities, some originating from Congress and some from the Commission. Space does not permit a thorough exposition of these rules, especially since some content regulations, like those involving ownership limitations, are in a constant state of flux. Some issues, however, have been more enduring than others, and we turn now to a brief examination of several of these.

Children's programming

The emotional and cultural welfare of children is a continuing concern for American society's moral and legal gatekeepers, and this obsession has been reflected in the federal government's supervision of the broadcast programming marketplace. The formative days of television were characterized by a plethora of children's programs. They were a staple of the local broadcast day. But economic conditions forced broadcasters to abandon most local production, except for news, and the task of meeting the needs of children fell to the networks. They responded with blocks of animated, made-for-TV series aired primarily on Saturday mornings. In 1974, the FCC issued a policy statement exhorting licensees to increase the amount of children's programming, recommending that it contain more informational and educational content, and expressing concern over the commercial practices employed in some children's shows, such as cartoon characters also being featured in commercials in and around the program. The Commission eventually concluded that the policy statement had been ineffectual, and in the spirit of deregulation during the administration of President Ronald Reagan, the agency made a tactical decision to leave the matter of children's programming to the vagaries of the marketplace. The FCC attempted to deflect criticism of its *laissez-faire* attitude towards the youngest audience demographic by arguing that the emergence of new technologies, such as cable and VCRs, would help to meet the needs of children.[32]

The Commission's deregulatory approach, particularly as it related to commercial practices in children's programming, was greeted with undisguised skepticism in the federal courts,[33] prompting a reconsideration of the agency's hands-off approach to children's programming and commercial standards. Under pressure from parents and consumer groups, Congress took matters into its own hands and in 1991 enacted a series of specific children's programming requirements. The FCC subsequently adopted rules to implement the Congressional mandate. These rules, which also apply to cable television, limit commercial time in children's programming produced primarily for viewers 12 years or younger to 10.5 minutes per hour on weekends and 12 minutes per hour on weekdays. The FCC also prohibits program length commercials (programs associated with a product, in which "commercials for that product are aired").[34] In addition, commercial TV stations are required to air some programming that meets the educational and informational needs of children and must also document these efforts in the public file. In 1997, the FCC expanded their rules to require broadcast stations (but not cable systems) to devote three hours per week to programming specifically designed to meet the educational and information needs of children 16 and younger.

Obscenity and indecency

Near the close of CBS's telecast of the 2004 Super Bowl half-time show, millions of shocked fans watched as pop star Justin Timberlake ripped off part of singer Janet Jackson's corset, exposing her right breast. FCC Chairman Michael Powell was shocked, too. Calling the stunt "classless, crass and deplorable," he promised an immediate investigation.[35] For violating the Commission's rules against indecency, the network's parent company, Viacom, was fined $550,000 representing $27,500 (the maximum allowed for a single violation) for each of the 20 affiliates owned by the network.[36] This incident may have been unusual for the publicity it generated, but the Commission's concern with regulating indecency on the airwaves has its genesis in the permissive decade of the 1960s. In 1964, when their licenses came up for renewal, the FCC considered complaints against several Pacifica Foundation stations for programming that included, among other things, a candid discussion by eight homosexuals about their problems and lifestyle. The Commission rejected the complaints on free speech grounds[37] but as the rhetoric on radio became increasingly expletive-laden, the agency's patience ran out and it began handing out fines.

The legal authority for the government to regulate obscene and indecent content is contained in Title 18, Section 1464 of the US

Criminal Code, which declares unambiguously: "Whoever utters any obscene, indecent, or profane language by means of radio communication shall be fined under this title or imprisoned not more than two years, or both."[38] These criminal sanctions are supplemented by the FCC's administrative authority to order a broadcaster to cease and desist from airing indecent or obscene material.[39] Obscenity, sometimes referred to as hard core pornography, has not been much of an issue for over-the-air broadcasters. The Supreme Court has declared obscenity to be devoid of First Amendment protection because of its lack of social value.[40] Indecency, on the other hand, does enjoy some constitutional protection as a class of speech. Nevertheless, the FCC has placed restrictions on the *broadcast* of indecent material which, according to the Commission, includes "language that describes, in terms patently offensive as measured by contemporary community standards for the broadcast medium, sexual or excretory activities and organs, *at times when there is a reasonable risk children may be in the audience.*"[41]

In 1978, in *FCC v. Pacifica Foundation*,[42] the Supreme Court upheld an indecency complaint against a New York radio station for its broadcast of a 12-minute George Carlin monologue, taken from one of Carlin's albums, in which Carlin discussed the "seven dirty words you can never say on TV" (see Figure 4.2). Although Section 326 of the Communications Act (the no censorship provision) precludes government editing of material prior to airing, this ban does not "deny the FCC the power to review the content of completed broadcasts" the Court declared, or "limit the FCC's authority to sanction licensees who engage in obscene, indecent, or profane broadcasting."[43] The Court also observed that, of all forms of communication, broadcasting has the most limited First Amendment protection. Broadcasting is a uniquely pervasive medium in the lives of people, extending into the privacy of the home, and is "uniquely accessible to children." Thus, the Court accorded the FCC greater latitude in regulating content that is patently offensive.[44]

Several years later the Commission extended its indecency policy beyond Carlin's "seven dirty words" but also created a "safe harbor" between midnight and 6 a.m. (later modified under court order to 10 p.m. to 6 a.m.), when broadcasters could air programs that might otherwise be deemed indecent. A Congressional attempt to impose a 24-hour indecency ban ran afoul of the Court of Appeals for the District of Columbia and was declared unconstitutional.[45]

Figure 4.2 Comedian George Carlin became an involuntary flashpoint for indecency regulation when WBAI aired one of his album monologues. Source: Photo courtesy of Carlin Productions.

In recent years, as the Washington political and regulatory climate has grown increasingly conservative, the FCC has aggressively sanctioned stations for airing indecent content, and some of the fines have been substantial. For their part, broadcasters have complained that the ad hoc nature of much of this decision-making resulted in Commission indecency standards that have become a confusing and disturbingly vague morass.

As a consequence, the FCC in 2001 released a policy statement "to provide guidance to the broadcasting industry" regarding the agency's indecency enforcement procedures.[46] The statement reaffirmed its holding in *Pacifica* that indecency findings involve at least two fundamental determinations – that is, "the material must describe or depict sexual or excretory organs and activities" and must be "*patently offensive* as measured by contemporary community standards for the broadcast medium."[47] The Commission then set forth three principal factors that had been pivotal in the Commission's indecency decisions to date: "(1) the explicitness or graphic nature of the description or depiction of sexual or excretory organs or activities; (2) whether the material dwells or repeats at length descriptions of sexual or excretory organs or activities; (3) whether the material appears to pander or is used to titillate, or whether the material appears to have been presented for its shock value."[48] As further clarification of its indecency standards, the Commission then provided a litany of concrete examples from past cases.

The Janet Jackson episode described at the outset of this section arguably did not meet these standards, but the Super Bowl debacle sent an unmistakable shiver throughout the broadcasting industry. Television broadcasters began a much more thorough review of program content, erring on the side of caution when in any doubt. For example, on Veterans' Day in November 2004, following several complaints, 66 ABC stations decided against showing the award-winning film *Saving Private Ryan* for fear of running afoul of the FCC's indecency standards.[49] The film, which had aired twice previously on the network without complaint or the imposition of fines, was a realistic portrayal of the US Normandy landing in World War II and contained numerous incidents of profanity and violence. In February 2005 the Commission ruled unanimously that the ABC affiliates who did choose to broadcast the program did not violate its indecency rules.[50]

Violence on television

The media stand accused in the court of public opinion of cultivating a more violent society through a relentless barrage of violent entertainment. This complaint pre-dates the advent of television but the medium's pervasive influence on so many lives has made it an easy target for consumer groups and other critics who are convinced of the industry's complicity in creating a more violent cultural environment. The government's authority to regulate violent content directly, however, is constitutionally much more limited than in the case of obscenity and indecency. Therefore, under pressure from conservative watchdog groups, Congress has taken a more indirect approach. In the Telecommunications Act of 1996, Congress included a requirement that TV set manufacturers install a microchip – commonly referred to as the *V-chip* – in all sets 13 inches or larger manufactured after January 1, 2000. This chip allows viewers (e.g., parents) to block out programs with a certain rating. The rating system, which was devised by the FCC and is subscribed to by most television licensees and cable operators, is patterned after the movie rating system, although the categories are different.

Contests, hoaxes and lotteries

Radio and television are still the most pervasive entertainment media and competition and market pressures generate a panoply of promotional devices. Most of these are a form of rhetorical hyperbole that pose no legal problems. But the FCC draws the line at program content that is misleading, deceptive, or false. Some stations, for example, use contests as promotional tools. A contest is a plan in which a prize is offered or awarded to the public based upon chance, diligence, knowledge, or skill.[51] The broadcast of contest information is legal as long the station fully discloses the material terms of the contest and conducts the contest substantially as announced or advertised. No contest description may be false, misleading, or deceptive with respect to any material aspect.

The FCC is particularly wary of hoaxes in which the deliberate broadcast of false information is contrary to the public's welfare. A case in point is the 1990 broadcast by KROQ-FM (Pasadena, California) of a false murder confession from a caller who was actually a friend of the morning air personalities. Local police were not amused. The following year a Missouri station aired a phony distress signal in connection with a mock civil defense warning issued by its morning announcer. In still another case, a station sent out an erroneous report that one of its employees had been shot while on duty.[52] Although the FCC has always taken a dim view of hoaxes, incidents such as these prompted the Commission in 1992 to issue formal prohibitions against transmitting false information about a crime or catastrophe when it is foreseeable that the broadcast of such information will cause substantial public harm and such harm does in fact occur. However, the hoax rule exempts from its coverage any program accompanied by a proper disclaimer that the program is fiction if the disclaimer is "presented in a way that is reasonable under the circumstances."[53]

Both federal law and FCC rules prohibit the advertising and promotion of lotteries, including casino gambling. A lottery consists of three elements: chance, consideration, and a prize. The element of chance is present if no skill is required to win. Consideration consists of exchanging something of value (e.g., money) to win or participate in a game. A prize is anything of value offered to a participant, which may include such things as money, services, refund, or a gift. Because of gambling's allegedly pernicious influence on society, Congress and the FCC have historically taken a paternalistic role in monitoring the broadcast industry's contributions to this cultural malaise. However, two trends have emerged to narrow the scope of the government's regulation of the broadcast promotion of lotteries and other forms of gambling activities.

First, many states have sanctioned gambling, particularly casinos, as a much-need additional revenue stream. This has undercut the FCC's claim of gambling's demoralizing influence on society. Second, Congress and the courts have carved out numerous exceptions to the federal prohibition against the advertising of lotteries and other games of chance. These include such activities as state-sponsored lotteries, certain forms of fishing contests, not-for-profit lotteries, and gaming on Native American lands. Further, in 1999 the Supreme Court held that the federal government cannot enforce its ban against the advertising of casino gambling in those states where such activity is legal.[54]

Audience deception: quiz show rigging and payola

In the early 1950s the desire for non-controversial but popular entertainment nudged the television networks towards programming that capitalized upon human weaknesses such as greed and morbidity. Giveaway programs provided manufacturers with free promotional time without having to purchase ads, and they appealed to viewers' materialistic instincts. Such shows were also ratings bonanzas.

By the end of the decade, prime time hits like *Strike It Rich*, *The Sixty-Four Thousand Dollar Question*, and *Twenty-One* were attracting millions of viewers who had a morbid fascination with "ordinary" contestants who struggled each week with a parade of challenging questions. As broadcast scholar Sydney Head describes the fascinating spectacle:

> In dramatic confrontations between chorus-girl experts on astronomy, minister experts on love stories, shoemaker experts on opera, college-teacher experts on everything, contestants won and lost hundreds of thousands of dollars in a single night before the television cameras. Authenticity and drama were heightened by contestants caged in "isolation booths," armed guards and bank vice-presidents opening strong boxes on camera to removed sealed envelopes containing golden questions . . . Enboothed contestants raised the tensions still higher with lip-biting, brow-wrinkling, eye-rolling histrionics.[55]

There was just one catch. Such histrionics were neither natural nor spontaneous and the contestants were not quite as knowledgeable as they seemed. *The quiz shows were rigged!* Contestants were frequently provided answers in advance, they knew in advance what to expect, and they were coached on how to behave for maximum emotional impact. Although rumors of quiz show "rigging" had surfaced as early as 1956, three years later the television industry found itself in the middle of a full-blown scandal propelled by government investigation that featured televised Congressional hearings. Energized by what they viewed as a massive fraud perpetuated upon the American public, legislators threatened to bring the networks directly under government control. Instead, cooler heads prevailed, and Congress amended the Communications Act to prohibit the offering of any assistance to program contestants with the intent to deceive the audience. To this day, quiz shows that provide any assistance to participants, such as sample questions, acknowledge this fact in a disclaimer to avoid accusations of audience deception.

Another practice that attracted Congressional attention in the late 1950s and early 1960s was *payola* – payments (bribes) by record distributors to disc jockeys for favored treatment of their latest releases. Popular DJ's controlled the playlists for their stations, and the inclusion of a new release on this list could significantly affect a record's success. Thus, competitive pressures within the record industry led distributors to seek favored treatment for their artists through direct payments to station deejays of money or other gratuities. Meanwhile, listeners remained blissfully ignorant that their artistic options were being subtly manipulated. Following an investigation, Congress again amended the Communications Act to provide that "any employee of a radio station who accepts or agrees to accept from any person . . . who pays or agrees to pay such employee, any money, service or other valuable consideration for the broadcast of any matter over such station shall, in advance of such broadcast, disclose the fact of such acceptance or agreement to such station."[56] What is interesting is that Congress chose not to prohibit the practice entirely. Instead, it simply required disclosure to station management when an employee does accept money or other consideration for providing preferred treatment to a record company. Nevertheless, the practice of payola still occasionally generates headlines. This should not be surprising considering the symbiotic relationship of the recording and radio industries. Recorded music is the lifeblood of radio stations, and publicity generated through the broadcast of new releases is crucial to the recording industry's bottom line.

Political Programming and the Public Sphere

The evolution of broadcasting as a mass medium posed new challenges for democratic discourse. Candidates could now transmit political messages simultaneously into the homes of thousands or even millions of listeners and viewers, creating an unprecedented sense of political community. Yet, the traditional theory of republican government had assumed the centrality of the printed press as a medium of political expression. (This presupposed, of course, a press that was free from government control.) But the broadcasting "press" was burdened by a natural limitation, the scarcity of radio waves, which provided a rationale for greater government oversight than was the case with the print media. In addition, whether stations and networks were government owned or administered (as in Europe and other parts of the world), or privately owned (as in the United States), control of radio in its formative stages was far more concentrated than control of the traditional printed press. Thus, both authoritarian and democratic nations recognized that there was a risk that "people in power would use their sway over the command centers of broadcasting to deny the opposition an equal opportunity to disseminate its views and compete for votes."[57]

In the United States, radio quickly became an advertiser-supported medium. This posed a number of challenges for government regulators. Because advertising was expensive, what safeguards could be put in place to ensure that the best-financed candidates did not monopolize airtime? Should candidates be guaranteed free time, or should access depend entirely upon their financial resources? Should all candidates for a given office be guaranteed equal time? What would prevent an extremist politician from unduly influencing public opinion? If candidates were guaranteed airtime, what role should broadcasters play in monitoring the content of political broadcasts?

Countries where broadcasting was government controlled, as in Europe, responded to these concerns by allocating free airtime to parties, rather than individual candidates (this model of political access is still prevalent in Europe). The American response, on the other hand, reflected the highly personal nature of its political system in which individual allure and charisma often trump party platforms.

The public policy concerning broadcast access for political candidates was first codified in the Radio Act of 1927 and then transported into its more durable successor, the Communications Act of 1934. Section 315 of the Act now provides, in part:

> If any licensee shall permit any person who is a legally qualified candidate for any public office to use a broadcasting station, he shall afford equal opportunities to all other such candidates for that office in the use of such broadcasting station. *Provided*, that such licensee shall have no power of censorship over the material broadcast under the provisions of this section. No obligation is imposed under this subsection upon any licensee to allow the use of its station by any such candidate.

Thus, when a station (cable operators who originate political broadcasts are also bound by Section 315) permits a legally qualified candidate for a given office to use its facilities, all opposing candidates for that office are entitled to "equal opportunity," that is to say equal treatment, which includes but is not limited to the opportunity to purchase equal amounts of air time. Station personnel are strictly prohibited from censoring a candidate's message no matter how offensive or outrageous it might be.

In addition, Section 315 does not require licensees to provide time for candidates for *all* offices. They can ignore races, for example, that might not be of interest to their listeners or viewers. However, there is one notable exception to this principle. Under an amendment to the Communications Act enacted in 1971 (Section 312), licensees must allow *reasonable access* to candidates for *federal elective office*. Understandably, broadcasters are frustrated by such vague standards as *reasonable access* and are at the mercy of the FCC's rather cumbersome ad hoc decision-making process in comprehending their obligations under federal law.[58]

To prevent an undue burden on a station's journalistic mission during political campaigns, in 1959 Congress amended Section 315 to exempt certain types of news programs from its coverage. Thus, appearances by political candidates on bona fide newscasts, news interviews, news documentaries, or on-the-spot coverage of news events, including broadcast debates and press conferences, do not obligate stations to provide equal opportunities to opposing candidates.

Other on-air appearances by candidates, even if not politically motivated or presented within an entertainment format, usually do fall under Section 315. Celebrities who suddenly acquire political ambitions can be particularly burdensome for licensees because appearances in entertainment fare do constitute a "use" under Section 315. A case in point is the 2003 gubernatorial campaign of Arnold Schwarzenegger during which the televising of old Schwarzenegger movies such as *Total Recall* or *Conan the Barbarian* could have prompted requests from the more than 100 candidates who had qualified for this special recall election.[59]

Under Section 315, political candidates are given breaks on advertising rates during certain periods prior to an election. Concerned about the spiraling cost of campaigning, Congress in 1971 amended Section 315 to require that 45 days prior to a primary and 60 days before a general election broadcasters provide air time to candidates at the "lowest unit rate" (LUR) Not only may licensees *not* inflate the commercial costs to candidates but they must charge candidates the same rate their most favored advertisers pay for the same length of commercials and for the same time of day (such as television prime time or radio drive time). Whereas commercial advertisers must buy a certain quantity of spots to qualify for discounts, political candidates during the period described above are entitled to these discounted rates regardless of how few spots they purchase.[60]

All candidates are entitled to the lowest unit rate, but in the Bipartisan Campaign Reform Act of 2002 Congress added an additional requirement for federal candidates. To qualify for the LUR, federal candidates must also identify themselves at the end of their broadcasts and explicitly state that they approved the message. Television spots must also include a photograph or similar image of the candidate.

The fairness doctrine and the discussion of public issues

From its beginning, radio broadcasting has been afflicted with second-class First Amendment status. As early as 1929, the Federal Radio Commission reminded broadcasters of their public interest obligations, including the presentation of controversial issues of public importance:

> It would not be fair, indeed it would not be good service to the public to allow a one-sided presentation of the political issues of a campaign. In so far as a program consists of discussion of public questions, public interest requires ample play for the free and fair competition of opposing views, and the commission believes that the principle applies not only to addresses by political candidates but to all discussions of issues of importance to the public.[61]

No such requirement would have withstood constitutional scrutiny if applied to the printed press. Further, in 1941, the Federal Communications Commission held that a licensee could not be an advocate for a partisan cause, that is, it could not editorialize.[62] This denied broadcasters a privilege that was constitutionally sanctified for their print brethren. Conversely, the Commission reaffirmed a broadcaster's legal *duty* to afford time for the discussion of public issues:

> Freedom of speech on the radio must be broad enough to provide full and equal opportunity for the presentation to the public of all sides of public issues. Indeed, as one licensed to operate in a public domain the licensee has assumed the obligation of presenting all sides of important public issues, fairly, objectively and without bias. The public interest — not the private — is paramount.[63]

In 1949, the FCC abandoned its policy against broadcast editorials, declaring that licensees could editorialize provided that they provide opportunities for opposing points of view. This regulatory change of heart was outlined in some detail in a document formally titled *In the Matter of Editorializing by Broadcast Licensees*,[64] which became the basis of the Commission's *fairness doctrine* that remained in effect until it was repealed almost 40 years later. As originally conceptualized in its editorializing report, the Commission imposed a two-fold obligation upon broadcasters: to devote a reasonable amount of time to the discussion of controversial issues of public importance *and* to provide this coverage in a balanced and fair manner. However, this seemingly simple directive masked the doctrine's operational complexity as it evolved under the Commission's tutelage.

In the 1960s the FCC embellished the fairness doctrine with the addition of two corollary doctrines, the Personal Attack Rules and the Political Editorializing Rules. The Personal Attack Rules required broadcasters to provide free reply time to anyone whose honesty or integrity was impugned during the discussion of a controversial issue of public importance. The rules did not apply to such attacks made during a bona fide news program. The Political Editorializing Rules provided that, if a licensee editorially endorsed or opposed a political candidate, it was obligated to provide free air time to representatives of opposing candidates to respond to the station's editorial. This understandably dampened broadcasters' enthusiasm for political engagement and, along with the requirements of the fairness doctrine itself, buttressed their complaint of second-class First Amendment status.

In 1969, a unanimous Supreme Court, in *Red Lion Broadcasting Co., Inc. v. Federal Communications Commission*, affirmed the federal government's authority to require broadcasters to provide their audiences with a balanced presentation of controversial issues of public importance.[65] The Court was clearly unimpressed with broadcasters' argument that they deserved Constitutional parity with their print counterparts:

> [A]s far as the First Amendment is concerned those who are licensed stand no better than those to whom licenses are refused. A license permits broadcasting, but the licensee has no constitutional right to be the one who holds the license or to monopolize a radio frequency to the exclusion of his fellow citizens. There is nothing in the First Amendment which prevents the Government from requiring a licensee

to share his frequency with others and to conduct himself as a proxy or fiduciary with obligations to present those views and voices which are representative of his community and which would otherwise, by necessity, be barred from the airwaves.

Because of the scarcity of radio frequencies, the Government is permitted to put restraints on licensees in favor of others whose views should be expressed on this unique medium. . . . It is the right of the viewers and listeners, not the right of the broadcasters, which is paramount.[66]

The *Red Lion* decision was a disappointing defeat for broadcasters. And their temperament was not brightened by the Court's declaration just five years later that a state law requiring a newspaper to provide reply space for a political candidate who had been editorially criticized was unconstitutional.[67] In 1987, however, a deregulation-minded FCC repealed the fairness doctrine after it concluded that the policy was unconstitutional after all. The Commission reached this conclusion based upon doubts about whether spectrum scarcity was still an issue in a new technological age and whether the doctrine was continuing to serve the public interest. The Personal Attack and Political Editorializing rules remained in effect for more than a decade until a federal appeals court in 2000 ordered the Commission to abandon these rules as well.[68] While the elimination of the fairness doctrine appears to be a long-awaited vindication for broadcasters, they are painfully aware that no federal court has declared the doctrine unconstitutional and that it could still be reinstituted by Congress or the FCC.

Policing New Technologies

Cable television

Cable television is so ubiquitous that to refer to it as a "new" technology might appear anachronistic. Nevertheless, it dramatically reconfigured the nature of the traditional television industry, and its potential is still evolving.

Originally known as CATV or *Community Antenna Television*, cable television was developed in the late 1940s to serve communities unable to receive over-the-air TV signals because of terrain or geographical remoteness from transmitters. But over time it became apparent that cable television would also become a major competitor to existing TV stations by offering program content of its own. Cable expanded to promote the potential of endless channel capacity, interactive communication, and other technological novelties. However, the realization of these promises depended upon a level regulatory playing field for cable and broadcast interests alike.

Influenced by the power of the broadcasting industry, the FCC's instinctive reaction to the budding cable technology was to protect the interests of local broadcast stations. Since cable systems did not use the electromagnetic spectrum directly, they were not licensed by the Commission. But the agency did license the microwave relay services that delivered the additional cable channels required to attract subscribers in the larger markets. In so doing, the FCC took into account the economic impact on local television stations of licensing a microwave relay service in their communities and whether authorizing such a relay system would adversely affect local stations' capacity to provide effective broadcast service. In 1963, a federal appeals court affirmed the Commission's power to consider the effect on existing local stations of licensing microwave services for cable.[69] Three years later an expansive Commission established rules for *all* cable systems regardless of whether or not they were served by microwave; a regulatory initiative that was affirmed by the Supreme Court.[70] The rules, which institutionalized the agency's protectionist attitude towards the broadcasting industry, required that cable systems carry the signals of local broadcast stations (known as the "*must carry*" rule) and prohibited systems from importing distant stations that duplicated the programming of local stations (the "*nonduplication*" rule).[71] The must carry rules would become increasingly controversial as the number of channels and the availability of cable programming grew. About the same time the FCC announced a freeze on any authorizations of microwave systems in the "top 100" broadcast markets until the agency could adopt a more comprehensive set of cable industry rules. This freeze remained in force for more than seven years, "effectively denying the cable TV industry any growth opportunities in the nation's most profitable broadcast markets from 1965 to 1972."[72]

Cable's growing popularity in the late 1960s convinced the Commission that the industry must serve the public interest in some way. In 1969, the agency decided that systems with more than

3500 subscribers must originate programming and provide facilities for production. Many cable operators took exception to what they viewed as unauthorized governmental intrusion, and in *U.S. v. Midwest Video*,[73] a cable company challenged these programming rules. However, in a 5–4 decision the Supreme Court affirmed the Commission's authority to regulate cable in the public interest because the origination rule would "further the achievement of long-established regulatory goals in the field of television broadcasting by increasing the number of outlets for community self-expression and augmenting the public's choice of programs and types of services."[74]

For the next decade the FCC continued its policy of insulating the broadcast industry from the competitive pressures of cable. For example, the Commission established franchising standards and required operators in the top 100 markets to provide their subscribers with a minimum of 20 channels, two-way capability, public access channels, and program origination. In addition, those cable systems that did originate programming were subject to many of the same programming requirements as broadcasters, such as the fairness doctrine, equal opportunity obligations, and sponsor identification regulations. Perhaps the most egregious example of the FCC's broadcaster-friendly policy was its adoption of anti-siphoning rules in 1975 devised to prevent cable systems from taking sports and movie programming away from broadcasters. But two years later a federal court ruled that the Commission had exceeded its authority,[75] and the rules were repealed.

As the number of cable channels and available programming proliferated, the FCC's *must carry* rules, which were still in effect, became increasingly burdensome for cable operators. From the cable industry's perspective, two constitutional claims were pertinent to their increasing discomfort with the rules: (1) the *must carry* obligation could be construed as a violation of cable operators' First Amendment free speech rights as well as (2) triggering an unconstitutional taking of private property.

In 1980, Turner Broadcasting System petitioned the Commission to rescind these rules while another system, Quincy Cable TV, took matters into its own hands by dropping two station signals without FCC approval, arguing that these broadcasters did not carry programming of interest to the public. The Commission denied Turner's request and fined Quincy $5000 for its action. Both appealed to the US Court of Appeals for the District of Columbia, which held the rules to be in violation of the First Amendment and outside the scope of the agency's regulatory mandate.[76] The Commission then promulgated a set of interim must carry rules, which were also declared unconstitutional.[77]

During the 1980s and 1990s, Congress became more actively involved in the cable industry. In the Cable Communications Policy Act of 1984, for example, Congress recognized the enormous economic investment required to install and maintain a cable system. Therefore it provided operators with an expectation of franchise renewal by local governments unless these governmental units could demonstrate that the franchisee had failed to meet their communities' needs. The statute also capped the fee that cities could charge for the right to use public streets and telephone poles. In addition, communities were prohibited from demanding that cable systems carry particular program offerings. The 1984 Act was soon followed by an increase in consumer dissatisfaction about what they perceived as unjustified rate increases and substandard service.[78] Congress responded with the Cable Television Consumer Protection and Competition Act of 1992, which re-imposed FCC and local governmental authority to regulate rates. It also required the Commission to establish standards of minimum service. Four years later, however, in the Telecommunications Act of 1996, Congress eliminated the rate regulations, which once again resulted in immediate increases and customer complaints. To foster competition, the Act also eliminated most long-standing barriers to competition between cable systems and phone companies as each sought to offer the other's services.

In the 1990s, broadcast signal carriage remained a bone of contention between broadcasters and cable operators. From the outset, broadcasters had argued they were entitled to compensation from cable owners for cable's carriage of their signals. They viewed this as a matter of simple fairness since cable companies regularly paid for cable networks such as CNN, Discovery, ESPN, and MTV. Cable operators responded that their carriage helped broadcasters by providing them with a larger audience and a cleaner signal.[79] In the 1992 Act, Congress attempted to reconcile these competing claims by providing two options for broadcasters who wished to remain on the cable. The first option was the traditional "must carry" under which the cable company would be required to place the local station on its basic tier – but would pay the broadcaster nothing. Alternatively, if the station believed its signal was valuable enough to merit financial compensation, it could select the second option.

Referred to as *retransmission consent*, this option involved the cable system paying the broadcaster a mutually agreed-upon fee in exchange for approval to carry that broadcaster's signal. If the negotiations failed, then the cable company would be prohibited from retransmitting the station. In 1997, the Supreme Court examined these new must carry rules and found them to be constitutional.[80]

Two other rules exemplify the FCC's continuing concern with protecting traditional station licensees from cable's economic threat. The *Network Nonduplication Rules* prohibit cable systems with more than 1000 subscribers from carrying distant network affiliates if they already carry a local affiliate of that network. The rules are predicated upon the marketplace reality that if subscribers are provided with a choice of two identical network affiliates they will divide their choices, resulting in fewer viewers for each affiliate and hence less advertising revenue. Meanwhile, under the *Syndicated Exclusivity* ("*Syndex*") *Rules*, local stations with exclusive rights to a syndicated program can demand that cable systems black out the show from being carried into their market via either a distant station or nationally broadcast superstation. (A similar rule was in effect prior to the 1980s but was abolished in the deregulation of the 1980s. "Syndex" was reinstated in 1990.)

The advent of digital technology, which made possible High Definition Television (HDTV) further complicated the already strained relations between broadcasters and cable operators. Because the technology for converting from the conventional analog television system to digital TV was expensive and the sheer magnitude of the conversion would be time-consuming, the FCC originally gave broadcast licensees until 2006 to complete the conversion. This deadline was later extended to 2009 after industry experts predicted that a majority of homes would not be equipped with digital sets by 2006. During the transition, of course, many stations would transmit both analog and digital signals. But the digitalization of the television industry resurrected the enduring tensions between cable operators and broadcasters over mandatory cable carriage of local TV stations. Would cable systems, for example, be required to carry the digital signals of broadcasters? If so, would they be required to carry both the analog and digital signals of a licensee during the transition period? And would they be required to carry the multiple digital signals that a single digital channel was capable of accommodating?

Broadcasters contended that must carry rules should be applied to digital TV to ensure its survival. The cable industry responded that such an extension of must carry obligations would violate the First Amendment rights of cable operators. In 2001, the FCC adopted rules addressing the issue of cable carriage of digital television. The Commission tentatively concluded, among other things, that requiring a cable system to carry both a broadcaster's analog *and* digital signals would impinge on cable operators' First Amendment rights. Instead, a commercial TV station broadcasting in both formats during the transition period could require a cable system to carry its analog signal through must carry and could negotiate the carriage of its digital signal through the retransmission consent provision described above. A station broadcasting only a digital signal could elect either must carry or retransmission consent.[81] Believing it was entitled to mandatory cable carriage of both digital and analog signals, an unsatisfied broadcasting industry petitioned the FCC for reconsideration of these rules, but in 2005 the Commission denied their plea.[82]

Direct broadcast satellite

The FCC approved its first *direct broadcast satellite* (DBS) licenses in 1982 against a backdrop of strong opposition from broadcasters who, like their cable counterparts, feared the potential competition that satellite communications promised. Broadcasters were understandably concerned about possible decline in their advertising revenues because DBS could import distant signals and thereby diminish the audiences for local stations. Broadcasters opposed the allocation of spectrum space for DBS and, as a minimum, petitioned the FCC to impose on DBS the same public interest requirements to which the broadcast industry was subjected under the Communications Act of 1934. The Commission declined to apply such a standard to DBS operators and left the industry virtually unregulated. DBS, in the FCC's view, could give rural viewers the same quality of television service long enjoyed by their urban counterparts, and the greater availability of channels would allow DBS operators to more effectively cater to all viewers' tastes.[83]

Broadcasters challenged the FCC's decision in court, arguing that satellite systems with a national footprint violated the Communications

Act's requirement of local licensing, deprived local TV stations of advertising revenue, and undermined programming directed at local interests.[84] The court generally supported the FCC's decision not to apply the public interest mandate to DBS operators.[85]

Despite the advances of DBS in the 1990s, until 1999 it could not provide subscribers with access to local television signals. This changed with the passage of the Satellite Home Viewer Improvement Act,[86] which provided that the carriage of one station in a market entitled all stations in the market to invoke the must carry or retransmission consent currently applied to cable.

Chapter Rewind

The history of the world's communications technologies reveals that they are usually conceived and cultivated in a politically unfettered environment but are eventually subjected to some form of regulation. Radio broadcasting, the world's first electronic mass medium, was no exception. In the United States, radio broadcasters in the 1920's battled each other on the electromagnetic spectrum, until the chaotic situation became untenable. The offenders themselves pleaded for government intervention to restore a sense of electronic stability to this fledgling industry. Congress responded in 1927 by creating the Federal Radio Commission followed seven years later by the Federal Communications Commission (FCC), which still regulates all forms of electronic communication.

Early in the Commission's tenure, broadcasters assumed that its authority would be limited to regulating frequency assignments and setting standards to prevent technical interference. But the FCC quickly broadened its regulatory sovereignty by declaring its intention to review the quality of a licensee's service in the public interest. This authority to review programming performance, as well as to license stations to ensure an efficient use of the electromagnetic spectrum, was sustained by the US Supreme Court.

The FCC was a product of the Roosevelt revolution during which the "regulatory state" was drastically expanded to deal with, among other things, a host of concerns surrounding the nation's business and industrial enterprises. However, unlike other businesses, radio and television are imbued with certain First Amendment entitlements. This reality is recognized in the Communications Act of 1934, which prohibits government censorship of the broadcast media. Unlike their print brethren, however, broadcasters' First Amendment interests are balanced against the public interest standard – an ambiguous mandate of the Communications Act that ultimately determines whether applicants are entitled to licenses and whether they get to keep them once they are issued.

The Commission has remained faithful to its "no censorship" restriction. Still, during its first four decades, the FCC erected a fairly elaborate regulatory enterprise that included, among other things, ownership limitations, programming standards and policies, and requirements that broadcasters provide air time for the discussion of controversial issues of public importance, referred to as the fairness doctrine. From the outset, the FCC has also enforced the Congressional mandate, incorporated into the Communications Act as Section 315, that political candidates be afforded equal opportunities to respond to their opponents' use of a licensee's facilities.

The "Reagan Revolution" of the 1980's altered the political landscape, and the Commission, reflecting the nation's conservative drift, embarked upon a program of deregulation. This was most obvious in the liberalization of the ownership caps, simplification of the licensing process, and the elimination of the fairness doctrine, which the FCC on its own initiative finally decided was unconstitutional.

In regulating the new electronic media the Commission has had to walk a thin line between a policy of protectionism for established media, such as radio and television broadcasting, and the need to provide breathing space for technological development. The agency's initial response to the advent of cable television, for example, was to erect a wall of economic protection around the existing broadcasting industry. However, despite the fact that cable is still subject to a variety of FCC rules, deregulation has resulted in unprecedented cable growth, consolidation of the cable industry, and continuing consumer complaints about unjustified rate increases. With the advent of direct broadcast satellites (DBS), the FCC, perhaps recalling the stifling effects of government regulation in the early days of cable, refused to impose such restraints on satellite operators and instead encouraged DBS competition with both the cable and broadcast industries.

SELF-INTERROGATION

1   What problem led to the first international regulation of radio?
2   Why did early broadcasters demand government regulation of their fledgling industry?
3   In the early days of radio, what three factors led to the lack of recognition of broadcasting as a constitutionally protected medium?
4   Compare the new FCC's view of its authority under the Communications Act of 1934 with the view of the broadcast industry. How did the Supreme Court decide this dispute in 1943?
5   What is the primary rationale for government regulation of broadcasting in the United States?
6   Describe the three primary functions of the FCC.
7   Explain the three kinds of ownership limits the FCC has promulgated for the broadcast industry.
8   Describe the measures the FCC implemented in the 1990s to require broadcasters to respond to the interests of children.
9   What two justifications did the Supreme Court provide in the *Pacifica* case for the government's regulation of broadcast indecency?
10   Define payola and explain why this is a matter of concern to Congress and the FCC.
11   Summarize the basic requirements of Section 315 of the Communications Act of 1934 as they pertain to political candidates.
12   In 1969, the Supreme Court upheld the now defunct fairness doctrine requiring broadcasters to provide time for opposing views on controversial issues of public importance. What was the Court's justification in doing so? Why did the FCC ultimately decide that the doctrine was unconstitutional?
13   The *must carry rules* are among the most contentious of the FCC's cable television regulations. How do these rules reflect the Commission's protectionist attitude towards the broadcast industry? What two constitutional arguments might be cited in support of the cable industry's opposition to *must carry*?
14   With the arrival of DBS, what arguments did the broadcast industry advance for regulating the satellite industry in the public interest?

NOTES

1   John D. Tomlinson, *The International Control of Radio Communications* (Ann Arbor, MI: J.W. Edwards, 1945), 11–12.
2   Madeleine Schachter, *Law of Internet Speech* (Durham, NC: Carolina Academic Press, 2001), 15.

3   Wireless Ship Act, ch. 379, 36 Stat. 629 (1910).

4   Schachter, *Law of Internet Speech*, 16.

5   Ch. 287, 37 Stat. 302 (1912).

6   Dwight L. Teeter, Jr. and Bill Loving, *Law of Mass Communication: Freedom and Control of Print and Broadcast Media*, 11th edn (New York: Foundation Press, 2004), 700.

7   Stuart Minor Benjamin, Douglas Gary Lichtman, and Howard A. Shelanski, *Telecommunications Law and Policy* (Durham, NC: Carolina Academic Press, 2001), 13.

8   Ibid.

9   Walter B. Emery, *Broadcasting and Government* (Lansing, MI: Michigan State University Press, 1961), 17.

10 United States v. Zenith Radio Corporation, 12 F. 614 (N.D. Ill., 1926).

11 35 Op. Attn. Gen. 126 (1926).

12 Sydney Head, *Broadcasting In America*, 2nd edn (Boston: Houghton Mifflin Co., 1972), 158.

13 Benjamin et al., *Telecommunications Law and Policy*, 17, quoting Clarence Dill, "A Traffic Cop for the Air," 75 Review of Reviews 181, 184 (1927).

14 Schachter, *Law of Internet Speech*, 16.

15 Trinity Methodist Church, South v. Federal Radio Commission, 47 F.2d 670 (C.A.D.C. 1931).

16 KFKB Broadcasting Association, KRC, June 13, 1930, affirmed 47 F.2d 670 (C.A.D.C. 1930).

17 Teeter and Lovomg, *Law of Mass Communication*, 708.

18 Don R. Pember and Clay Calvert, *Mass Media Law* (Boston, MA: McGraw-Hill, 2005), 586.

19 For example, see City of Los Angeles v. Preferred Communications, Inc., 476 U.S. 488, 496 (1986) (Blackmun, J., concurring); FCC v. League of Women Voters of California, 468 U.S. 364, 376 (1984); Film Mutual Corporation v. Industrial Commission of Ohio, 236 U.S. 230 (1915)

20 NBC v. U.S., 319 U.S. 190 (1943).

21 Teeter and Loving, *Law of Mass Communication*, 710. For additional rationales, see 710, note 32.

22 John D. Zelezny, *Communications Law: Liberties, Restraints, and the Modern Media*, 4th edn. (Belmont, CA: Wadsworth, 2004), 381.

23 Ibid, 371.

24 Head, *Broadcasting in America*, 185.

25 Ibid., 95.

26 Ibid., 200.

27 RCA v. U.S., 341 U.S. 412 (1951).

28 For example, see Fox Television Stations, Inc. v. FCC, 280 F.3d 1027 (D.C. Cir., 2002).

29   J. S. Adelstein, *Statement before the Senate Committee on Commerce, Science and Transportation hearing on FCC Oversight*, June 4, 2003, p. 2, retrieved August 2, 2005 from http://hraunfoss.fcc.gov/edocs_public/attach-match/DOC-235162A2.doc

30   Prometheus v. F.C.C., 373 F.3d 372 (3d Cir., 2004).

31   See Primer on Ascertainment of Community Needs, rev. 24 FCC 942, 39 FR 32288 (1974) (1971).

32   Kenneth C. Creech, *Electronic Media Law and Regulation*, 4th edn (Boston: Focal Press, 2003), 183–4.

33   See Action for Children's Television v. FCC, 821 F.2d 741 (1987).

34   "New Children's Television Rules Adopted," *Pike & Fischer's Broadcast Rules*, Service No. 27 (March/April 1991): 2.

35   "FCC to Investigate Super Bowl Breast-baring," CNN.com, February 2, 2004 (downloaded on February 2, 2004), located at http://cnn.usnews.printthis. See also Bill Carter, "Bracing for Fallout from Super Indignation," *The New York Times* (February 5, 2004), B1.

36   Notice of apparent liability for forfeiture, FCC 04–209, released September 22, 2004.

37   See In re Pacifica Foundation, 36 FCC 147 (January 22, 1964).

38   18 U.S.C.S. sec. 1464 (2005).

39   47 U.S.C. sec. 312(b), 503(b) (2000).

40   Roth v. United States, 354 U.S. 476 (1957).

41   56 FCC 2d 94, 98 (1975) (emphasis added)

42   438 U.S. 726 (1978).

43   Ibid., 735–8.

44   Ibid., 748–50.

45   Action for Children's Television v. Federal Communications Commission, 932 F.2d 1504 (1991).

46   *Industry Guidance on the Commission's Case Law Interpreting 18 U.S.C. sec. 1464 and Enforcement Policies Regarding Broadcast Indecency*, 16 F.C.C. Rcd 7999 (2001).

47   Ibid., sec. III.A.7–8.

48   Ibid., sec. III.B.10.

49   'Lisa de Morael, " 'Saving Private Ryan': A New Casualty of the Indecency War," *The Washington Post* (November 11, 2004), C01.

50   "In the Matter of Complaints Against Various Television Licensees Regarding Their Broadcast on November 11, 2004, of the ABC Television Network's Presentation of the Film 'Saving Private Ryan' ", File No. EB-04-IH-0589, February 3, 2005.

51   47 CFR, section 73.1216.

52   National Association of Broadcasters, *Contests, Lotteries & Casino Gambling* (Washington, DC, 1995), 24.

53   Ibid.

54   Greater New Orleans Broadcasting Ass'n v United States, 527 U.S. 173 (1999).
55   Head, *Broadcasting in America*, 339.
56   47 U.S.C. sec. 508(a) (2002)
57   Paul Starr, *The Creation of the Media* (New York: Basic Books, 2004), 370.
58   The Commission has identified four factors that should be taken into account: (1) how much time the candidate has already purchased; (2) the amount of disruption the purchase would create; (3) the potential for equal opportunities demands from other candidates; and (4) the timing of the request.
59   Roger L. Sadler, *Electronic Media Law* (Thousand Oaks, CA: Sage Publications, Inc., 2005), 48.
60   National Association of Broadcasters, *Political Broadcast Catechism*, 16th edn (Washington, DC, 2004), 33.
61   In Re Application of Great Lakes Broadcasting Co., 3 FRC Ann. Rep. 32 (1929).
62   The Mayflower Broadcasting Corporation, 8 FCC 333 (1940).
63   Ibid.
64   13 FCC 1246 (1949).
65   Red Lion Broadcasting Co., Inc. v. FCC, 395 U.S. 367 (1969)
66   Ibid, 390 (citations omitted).
67   Miami Herald v. Tornillo, 418 U.S. 241 (1974).
68   Radio-Television News Directors Association v. FCC, 229 F.2d 269 (D.C. Cir., 2000).
69   Carter Mountain Transmission Corp. v. FCC, 321 F.2d 359, 363 (D.C. Cir., 1963), cert. denied 375 U.S. 951 (1963).
70   United States v. Southwestern Cable Co., 392 U.S. 157 (1968).
71   See First Report and Order, 38 F.C.C. 683 (1965).
72   Teeter and Loving, *Law of Mass Communication*, 731.
73   406 U.S. 649 (1972).
74   Ibid., 668.
75   Home Box Office Inc. v. F.C.C., 567 F.2d 9 (D.C. Cir., 1977), cert. denied 434 U.S. 829 (1977).
76   Quincy Cable TV, Inc. v. F.C.C., 768 F.2d 1434 (D.C. Cir. 1985), cert. denied, 476 U.S. 1169 (1986).
77   Century Communications Corp. v. F.C.C., 835 F.2d 292 (D.C. Cir. 1987), cert. denied, 108 S.Ct. 2014 (1988).
78   Creech, *Electronic Media Law*, 93.
79   Ibid, 103.
80   Turner Broadcasting System v. FCC (II), 520 U.S. 180 (1997).
81   *Carriage of Digital Television Broadcast Signals: Amendment to Part 76 of the Commission's Rules*, etc., 16 FCC Rcd 2598 (2001).

82 *Carriage of Digital Television Broadcast Signals: Amendments to Part 76 of the Commission's Rules*, Second Report and Order and First Order on Reconsideration, CS Docket No. 98–120, FCC 05–27 (rel. Feb. 23, 2005)
83 Benjamin et al., *Telecommunications Law and Policy*, 542.
84 Ibid.
85 National Association of Broadcasters v. FCC, 740 F.2d 1190 (D.C. Cir. 1984).
86 113 Stat. 1501 (1999).

CHAPTER 5

Business Chronicles

W. Lawrence Patrick

Early radio station operators quickly realized the huge popularity of this new electronic medium. As department and electronics stores sold hundreds of thousands of receivers, families gathered by their living room radios to hear serial dramas, news and sports programming. Soon people were listening in their automobiles and later, with the advent of the transistor, they took their radios everywhere. Just as important, radio – the spoken word – made news and entertainment available to millions who did not have access to newspapers and magazines.

The Mass Audience

From President Roosevelt's "fireside chats" during the Depression to accounts of World War II brought nightly into American homes by war correspondents in Europe and the Pacific, radio became a medium that bound the nation together. It provided a common experience delivered instantly to millions across the nation and its comedies and variety shows brought respite from the world's scary realities. Likewise, when television made its way into so many American homes during the early 1950s, it also became an integral part of the culture. With early morning programs such as NBC's *Today* to prime time comedies, dramas and variety shows, as well as weekend sports broadcasts, television enmeshed itself in people's lives. This shift in information flow and entertainment delivery sparked the rise of something of great interest to advertisers – the *electronic mass audience.*

The electronic media allowed advertisers to instantly tell their stories and hawk their products to millions all across the nation. First radio, and then television, became advertisers' premier mass marketing tools. From toothpaste, automobiles, fast food and life insurance to political candidates, pharmaceutical products and luxury items, the electronic media sold all manner of goods, services and ideas to a truly massive (tonnage) audience. Advertising agencies and national advertisers increasingly appreciated the value of these expanding media and their ability to deliver millions of customers.

Once the temporary "freeze" on new television licenses was lifted by the Federal Communications Commission in 1952 and a national license allocation plan established, applications for new station licenses flooded the FCC. This was an electronic version of the nineteenth-century gold rush. The electronic media were transformed from an invention of engineers and hardware manufacturers to a business focused on attracting tonnage audiences and selling commercial sponsorships. For both radio and television stations, the 1960s through the 1990s were years of great profitability. Broadcasters saw that remote broadcasts from stores could draw thousands of consumers through the doors. Commercial time sales on stations skyrocketed and the value of broadcast properties increased dramatically. Stations were soon worth 10, 15 and even 20 times their annual profits. And many larger market stations were generating profit margins of 50 percent.

In the early years, advertisers sponsored entire programs. They created the scripts, produced the shows and sold their products within these shows. Later, when the costs of the programs grew too high for individual advertisers to support, stations and their networks began selling individual commercial messages called "*spots.*" These 30- and 60-second messages delivered audiences larger than could any newspaper or other advertising medium. And again, that delivery could be both instant and nationwide.

But advertisers needed to know how many and what type of people were listening to or viewing their commercials. Audience research companies, today led by A. C. Nielsen for television and Arbitron for radio, were created. These firms could track audiences and deliver demographic profiles of what type of listeners and viewers were tuned to particular stations and programs. Advertisers could then target their messages to reach the most desirable portion of an audience for their particular product or service.

Appealing programming was required to attract people to the advertising. But providing show content that consistently assembled large audiences became something beyond the scope of most local stations. Although some early radio outlets began to produce independent serial dramas such as *The Green Hornet* and *The Lone Ranger* as well as covering community events, most stations required outside resources for competitive, compelling on-air product. That programming need, and the willingness of national advertisers to satisfy it, created networks.

The Rise of the Network Model

Station owners quickly realized that *networks* could provide higher quality programming inexpensively. The networks could spread out the cost of high-priced talent and sports rights across hundreds of stations. They could also assemble a national audience that would attract deep-pocket national sponsors such as automobile manufacturers, food packagers, and beer and soft drink makers. On behalf of such advertisers, the radio and then, the television networks, developed orchestras, sports teams, high-profile journalists and former vaudeville stars as audience attractions. Linking stations together from studios in New York, the networks used such programming to simultaneously deliver a mass audience on behalf of participating sponsors.

NBC's Red and Blue Networks, as well as CBS, began by connecting radio stations via telephone lines. Outlets from Boston to Omaha to Albuquerque to Seattle could now simply throw a switch and re-transmit hundreds of hours of radio network programming every month. Local radio stations in even the smallest of towns could obtain national entertainment and news broadcasts by affiliating with a network. Later, with the advent of television, AT&T and smaller independent companies began to microwave video network programming from New York and Hollywood studios to stations across the country.

The networks were partners with local stations in the development of radio and television. They sold advertising within the programs and provided an incentive for all local stations to carry the shows simultaneously by paying station owners "compensation" – a fee for every hour of network programming the station broadcast. A major advertiser like Proctor & Gamble or Ford Motor Company would pay NBC or CBS for every commercial aired. Local stations would receive a small amount of that revenue in exchange for carrying the program. The amount of compensation paid to a station depended on the size of the market, the comparative strength of the station, and its commitment to carry the network schedule as broadcast. The networks expected simultaneous delivery of their programs and imbedded commercials and did not want local stations to pre-empt or delay the programming because this would disrupt the audience delivery promised to advertisers.

Networks recruited the best stage and screen performers to appear in programs. Writers, directors and talent were lured to radio and then television with the promise of lucrative contracts, instant national fame and steady employment. Many early radio and television stars also became linked with the advertisers who supported and promoted their programs. For instance, President Ronald Reagan, in his earlier career as an actor, was the spokesman for General Electric. Dinah Shore, a 1950s music and film star, became synonymous with Chevrolet, the sponsor of her weekly variety program.

Networks also possessed the resources to cover national and global news events. From election results to epic tragedies such as President Kennedy's death in Dallas to Neil Armstrong's first steps on the moon, television showed the American public the world beyond their local communities. The televised Nixon impeachment hearings during the Watergate scandal, plus coverage of the war in Vietnam, the civil rights movement and a multitude of conflicts on the other side of the world enabled the American public to see history being made first-hand. The images of great moments in sports and natural disasters such as hurricanes and floods were all beamed into households on a daily basis.

The broadcast networks also battled to develop dominant entertainment programming ranging from dramatic series such as *Dallas* or *CSI*, to comedies including *The Cosby Show, Seinfeld*, and *Everyone Loves Raymond*, to news broadcasts such as *60 Minutes* and *20/20*. The original three television networks, ABC, CBS and NBC, were extremely successful in delivering massive audiences for advertisers night after night. In the early 1990s, FOX joined the Big Three when Rupert Murdoch, an Australian media baron, bought a group of independent television stations from Metromedia and launched the fourth network. FOX's coup in luring away the rights to the National

Football League from CBS shortly thereafter marked FOX as a legitimate national television network.

In more recent years, smaller national networks, including the WB (owned by Time Warner), UPN (owned by CBS) and PAX (partially owned by NBC), as well as Spanish-language networks Telemundo (owned by NBC) and Univision joined the fight for national advertising dollars and viewers. In 2006, after enduring years of low audience delivery, the WB and UPN would be merged by their parent corporations into The CW network. Meanwhile, Trinity Broadcasting, with its major market owned-and-operated stations and a number of smaller market and low-power television outlets, became the first national religious network.

The *network/affiliate relationship* was mutually beneficial for many years. Networks produced or purchased programming and fed it to their affiliates. These local stations were happy to have most of their daytime and prime time schedules filled with network fare while also being paid compensation. Networks could afford to bid for increasingly expensive sports programming, such as major league and college football, baseball and basketball, while also producing specialty programming like the Olympics. The networks also created nightly news programs, supplied closed-circuit feeds of additional news and sports features that the stations could use in their local newscasts and covered breaking news at a level that individual stations could not afford to match.

The networks, however, faced a difficult dilemma. Operating a network and investing heavily in original on-air product for both daytime and prime time was a risky business. By the late 1980s, the now *four* major networks found that more than half of their new prime time series failed to gain sufficient audience levels to attract significant advertising dollars. When these shows failed, the tens or even hundreds of millions of dollars invested in them were wasted. This meant reduced network revenues. In recent years, networks have lost as much as $1 billion in advertising sales annually because of weak prime time content.[1]

In addition, the news operations, with their worldwide bureaus and thousands of producers, editors, writers and correspondents, became an increasingly large drain on network budgets. As cable began to build national audiences and Cable News Network presented its 24-hour constant news product, the traditional broadcast networks felt economic pressure. These networks were also in a transition period in which the original pioneers such as David Sarnoff at NBC and William Paley at CBS were succeeded by corporate owners – often from outside the broadcasting community. New owners like Loews Corporation's Lawrence Tisch who bought CBS, saw the television networks as just another commodity business so worked aggressively to reduce costs and increase profit margins without regard for broadcasting's unique public service role.

Meanwhile, sports rights, particularly at the professional level, sky-rocketed as team owners realized that television could cover their costs of operation and that the audiences for their events were vitally important to broadcasters and advertisers. Yet, despite the rising costs of sports and entertainment programming rights, network stockholders still demanded increased revenue and profit growth. Management was under intense pressure to deliver both ratings and revenues.

At first, wrenching decisions reducing the size and scope of the news operations were made. Hundreds of employees were dismissed. Bureaus were closed. Cuts reached into network management levels as well. Meanwhile, the pressure on programmers to develop hits became ever more intense. Network programming chiefs and many top executives were routinely dismissed for their inability to develop strong, ratings-attractive schedules. With so many dollars at stake, performance was critical and a bad program line-up could quickly sink an executive's career.

As another cost-saving step, the networks took aim at compensation payments to their affiliates.[2] Starting in the late 1990s, they made across-the-board compensation reductions. In cases such as the Lee Enterprise stations' struggle with NBC, compensation cuts led independent group owners to sell their outlets and get out of the business. By 2006, only a handful of stations received any compensation from a network, and these amounts were miniscule compared to prior years.

The networks also asked their affiliates to contribute to the rising cost of sports packages through either direct cash payments to the networks based on the size of their markets or by allowing the networks to retain more commercial time in sports programs. In some cases, the give-back was both money and additional time.

Paradoxically, while the networks themselves were under severe financial pressure, their *owned-and-operated* (O & O) groups were cash cows. From television's early days in the 1950s right up to the present, the network O & Os have always generated significant profits. For decades, these local stations in major markets were routinely generating 40, 50 or even 60 percent operating margins. Profits were tremendous. With between 12 and 20 local stations in the largest markets, the O & O divisions to a large extent subsidized their parent networks. The network companies realized that owning more local television stations was a good thing. National spot advertising at their 20 or 30 large-market local television stations could be offered to advertisers along with the network advertising packages. With strong local news operations that established their prominence with local viewers, these stations existed on a mix of strong national and local spot advertising. The O & Os became a solid cash flow machine that underwrote much of the networks' operations.

Many station groups that owned the rest of the network affiliates also understood how valuable local television stations were. These major companies such as COX, Gannett, Hearst-Argyle, LIN, Media General, Belo and dozens of others are either privately owned or, in some cases, large and diverse public media companies with platforms that include television, radio, newspapers, cable and other businesses. Since the early 1960s, these group owners have enjoyed decades of success by following a standard three-tactic formula for operating their television stations.

The *first tactic* for success was to be a traditional network affiliate and receive as much as 18–20 hours per day of programming delivered to the station by their network for broadcast to the local community. Granted, each network would routinely experience ratings increases and dips across the years. But local stations' financial performance consistently rode out the ups and downs of their network partners. The stations' base of strong local advertising sales financially insulated them from the effects of most of the network volatility. Local news revenues often accounted for nearly half of station income. This was something that the local station controlled and, as such, was a very valuable and stabilizing franchise. Annual cash flow margins in the medium and larger market stations were routinely in the 30–40 percent range and often higher. Smaller market stations generally did almost as well. For most of the past half century, it was difficult not to prosper as the owner of a local network affiliate. This was a franchise of enormous value with the potential to generate significant and steadily improving revenues and cash flow. Many station owners saw television as both a powerful force in their communities as well as an enormously lucrative business.

The *second tactic* for local station success was to focus ever greater effort on constantly honing their profitable *local news presence*. Originally viewed by some owners as a burdensome obligation imposed by government regulators, local news became the measure of financial success for most local stations by the early 1970's. By then, most evening newspapers in major metropolitan areas had died. Consumers were drawn to evening and late newscasts that could go "live" to breaking events and show them pictures of what was happening in their communities. News viewership rose and advertisers were attracted to the audience draw of this programming.

Local news today often generates from 35–50 percent of all revenues for a station because of its strong audience shares. At almost every affiliate, news is the largest department in terms of number of employees. The fight to promote a station's newscasts, to be "live, local and late breaking" and to control the rich avenues associated with being a market's #1-rated station in news is all-important for local outlets. News consultants, weather radar, helicopters and highly-paid talent are all weapons in the fight for local news supremacy.

The *third tactic* for local station success was to commit itself to negotiate the most favorable contracts possible in selecting and purchasing *syndicated programming*. Syndicated programming comes in two forms – first-run and off-network. First-run shows have not previously appeared on a network. They are developed by producers and sold to stations across the country – generally for airing in late afternoon or early evening prior to the network prime time hours. These programs can be on different network affiliates in different cities based solely on the negotiations with each local station owner. A show might appear on a CBS affiliate in Boston, an ABC affiliate in Tulsa and a FOX affiliate in Minneapolis. Among the most successful first-run properties are *Oprah, Dr. Phil, Wheel of Fortune, Jeopardy, Entertainment Tonight,* and *Judge Judy*. Stations generally make a multi-year commitment in purchasing them.

Conversely, off-network syndication programs are generally half-hour comedies that have previously been aired on one of the networks. These programs include *Seinfeld, Everybody Loves Raymond, The Drew Carey Show, King of Queens* and others. Usually, one-hour off-network dramas, such as *Law and Order* or the various *CSI* properties, are not an attractive purchase for local station operators because they play well only on late-nights and on weekends. Often busy with other household activities, audiences have not been willing to watch these 60-minute programs in the late afternoon or early evening periods where stations are most in need of syndicated fare. Some of these off-network hour dramas do find appropriate homes on basic cable channels, such as TBS, TNT, Court TV, and Bravo where they can even be scheduled in prime time.

The selection of syndicated programming is generally the most expensive decision that the local station manager will make. Shows such as *Oprah, Wheel of Fortune, Jeopardy, Entertainment Tonight, Seinfeld* and dozens of others are critically important to the station's well-being. They are enormously popular with viewers and, therefore, very attractive to advertisers. Syndicated shows' performance will usually affect the performance of programs that follow them in the schedule. *Oprah,* for example, has shown that it can deliver large audiences in late-afternoon. As a lead-in to a station's early newscast, *Oprah* can significantly improve the size of the newscast's audience. But such syndicated shows can also be very expensive for the station to buy.

Syndicators negotiate exclusive rights for these programs with local stations in multi-year packages worth millions of dollars per market. *Oprah,* for example, may cost $2.5 million or more a year in a top 10 market. Comedies, such as *Seinfeld* or *Everyone Loves Raymond,* may easily cost well over $1 million annually in many major markets. In smaller markets, these same programs may only cost $150,000 annually but this still represents a comparatively sizeable commitment for those station owners. Clearly, syndication is a real profit center for Hollywood television production companies.

With the relaxation of the FCC's ownership restrictions in the early 1990's, many affiliate groups began expanding through the purchase of additional stations. This made syndicated programs more affordable. The economies of scale of purchasing syndicated programming (as well as technical equipment, news wire service, and many other things) for 20 or more stations as compared to 5 or 6 was and is compelling.

A large group owner such as Raycom, Cox, LIN or Gannett, can make one deal for syndicated programming that prevents smaller groups or stand-alone operators in any of their markets from obtaining even an opportunity to purchase this programming. On the hardware side, a good example of economies of scale issue is seen in the shift to digital television transmission. By 2009, every television station has to shift to digital transmission. This requires purchasing many items, including a new transmitter. Although a single-station purchaser may pay $1.5 million for a transmitter, a company ordering 20 or more of these units may see their cost fall to only $1 million per unit. Similar savings are achieved in negotiating multi-station pacts with vendors such as Nielsen for ratings, the Associated Press for news service, and national sales representatives such as Katz as to the commission they will receive for selling local station time to national advertisers (national spot).

The networks' desire to add more stations to their O & O divisions, however, eventually set up a bitter confrontation with the affiliate groups. Networks began bidding for major market stations that traditionally were owned by major independent group owners. In some cases, such as the famous KRON-TV transaction in San Francisco, the NBC network canceled its network affiliation with the television station after losing the auction to purchase KRON-TV from its original owners. NBC subsequently bought another station in San Francisco, KNTV-TV, and switched its affiliation to that station – thereby putting the new owner of KRON-TV in an extremely precarious situation. This new owner had paid at least $500 million for an outlet that suddenly was an independent television station rather than a NBC affiliate and now had to secure every bit of its programming on its own.

Such aggressive network bidding on major market stations that were up for sale (with the threat of a loss of affiliation if the network did not win the bid), demands from the networks for "reverse" compensation (payments from the affiliates back to the network), and the networks' efforts to re-purpose some of their content by re-airing it on network-owned cable channels, all created tension with the affiliate groups. The affiliates filed lawsuits and FCC complaints against the networks. Efforts to contain the spread of network control through limiting their national reach were debated by Congress. By 2004, a compromise was hammered out. The networks' O & O station groups would be allowed to expand their reach from 35 to 39 percent of the total households in the United States. But this was less than the 45 percent they had sought. All four major networks are already near the 39 percent ceiling and are continuing their lobbying efforts to further raise it.

Rising sports fees, network/affiliate tension, and unyielding pressure to produce long-running hit prime time series were not the only dangers the networks came to face. The rise of cable television and other programming alternatives posed significant challenges to the networks and their mass audience model.

Cable Flexes Its Muscles

As discussed in Chapter 2, cable first appeared in rural areas as a way of providing basic television service to mountainous or sparsely populated regions. Pioneer cable operators like John Walson (see Figure 5.1) began importing distant station signals by building a master antenna on a nearby hillside and then feeding the signal by copper wire to individual homes. At first, broadcasters saw this as a benefit to their business because it often extended their signal to new consumers. However, cable soon grew to become a major competitor to broadcasting by offering alternative program services. As a consequence, regulatory and economic battles between the industries intensified.

Figure 5.1 John Walson overlooks Mahanoy City, Pennsylvania, the town he wired for cable in 1948. Source: Courtesy of Service Electric Cable TV, Inc.

When cable moved directly into cities where television stations were located, and cable systems imported distant signals of the "superstations" or cable network fare, broadcasters realized that these systems would compete for audiences and eventually for advertising revenues. Some broadcasters, like COX and the Post-Newsweek group, embraced the new competition by joining it. They invested in the cable industry, built or purchased local cable systems, and learned how this new medium worked. Most broadcasters, however, fought cable and attempted to stall its development through regulation.

"Must carry," compulsory copyright regulation, as well as network non-duplication and syndication protections were all efforts by broadcasters to stem cable's growing economic power. "Must carry" regulation shielded independent and smaller television stations from having a cable system transmit only the signals of the larger, network-affiliated television stations in their market. Compulsory copyright regulation required the cable companies to pay broadcasters, the movie studios and sports franchises for their programming. Network non-duplication and syndicated non-duplication protections meant that local cable systems could not import network or syndicated programs from distant television stations that duplicated the local station's offerings. This served to protect local station viewing levels and revenues. The Washington regulatory fights between cable and broadcast interests over these and other actual and proposed rules were both legendary and bitter.

By the early 1990s, when the cable industry had wired more than half the country and satellite distribution of video signals became both reliable and cost-effective, cable began a new chapter in its development. More and more national cable networks were launched. Cable News Network, MTV, Arts & Entertainment, ESPN and newer players all focused their programming on audience niches. Rather than attempting to assemble a mass audience with tens of millions of viewers nightly, these cable networks built franchises around smaller targeted audiences they could aggregate nationwide without competing head-to-head with the broadcast networks.

The growth of cable networks also created huge new avenues for video production. With their focus on nature and exploration programming, Discovery Networks generated thousands of hours of new content from independent producers. Music videos – the life-blood of MTV, VH-1, Country Music Television (CMT), Great American

County (GAC) and Black Entertainment Television (BET) – became an industry unto themselves. The growth of regional pay sports networks allowed professional sports teams, colleges and even minor sports leagues to find a video outlet for their games.[3] HGTV made home renovations and decorating a new video fixation. The Food Network filled homes with cooking recipes and preparation hints 24 hours a day.

The cable industry came to realize that consumers would pay for programming – and that one of the biggest demands was for movies. In 1975, HBO, a subsidiary of Time Warner, was the first to launch a national pay movie service. It satellite-delivered recently released movies to cable systems before they appeared on broadcast television. HBO was quickly followed by Showtime, Cinemax, Starz, American Movie Classics and other pay movie enterprises.

Cable programming expanded in other ways. While the broadcast networks clung to their Fall premieres and generally ordered only 22 original episodes per season (with many re-runs and specials to fill the extra weeks), cable moved in a different direction. The cable networks began launching original limited-run series against the networks' re-run schedule and aggressively promoted these edgier programs. *The Sopranos*, based on a real-life New Jersey crime family, *Six Feet Under, Nip/Tuck, The Larry David Show,* and *Deadwood* all drew significant cable audiences and critical acclaim. Because cable is free from broadcast indecency standards, these programs could afford to be more raw and realistic than traditional network television fare.

The broadcast networks did not totally ignore the success of cable programmers. In many cases, the cable networks were owned by the broadcast networks and broadcast group owners. ABC, for instance, purchased control of ESPN shortly after its launch. CBS was an early investor in Arts & Entertainment (A & E) and also launched MTV, VH-1 and Nickelodeon while also purchasing CMT, a country music channel, as well as Spike TV, an action channel focused on men.[4] NBC launched news and talk channels MSNBC and CNBC.

Group owners such as Scripps-Howard and Hearst Argyle also moved into national cable programming. Perhaps the most successful broadcast group to make this transition, Scripps created HGTV, the Food Network, the DIY network and the Fine Living network. It then purchased two other cable services – the Shop-at-Home channel and

GAC, a country music channel competitor to the CBS-owned CMT. For its part, Hearst became CBS's partner in the A & E channel.

The cable industry's move to replace its old copper-wire technology with a fiber optic delivery system, capable of carrying hundreds of channels, further fueled the proliferation of national cable networks. And it spelled the end of mass audiences for broadcast networks. Although still capable of delivering 50 million or more homes for special event telecasts such as the Super Bowl or the Oscar Awards, the broadcast networks now found themselves collectively reaching less than 50 percent of American viewers on a regular basis.

Advertisers first experimented with – and then enthusiastically embraced – cable networks as a way to reach targeted audiences for their products. Beer companies flocked to ESPN and its predominantly male viewers. Home improvement companies bought time on HGTV. And advertisers seeking children heavily supported both Nickelodeon and Toon Disney.

Fortunately, annual population growth and longer time spent viewing increased the overall size of the audiences for all video services. Advertising rates, even for the broadcast networks with their declining viewing shares, continued to rise throughout the 1990s. For many advertisers focused on reaching the largest possible number of consumers (a "tonnage" audience), broadcast networks still comprised an effective marketing medium. For automobile companies, soft drink manufacturers, fast food chains and dozens of other advertisers, network television remained the king. Nevertheless, these same advertisers also began diverting some of their advertising dollars to cable networks as a means of more precisely targeting their customer niches. For broadcast networks, it became an attack of a thousand cuts. No one cable network could equal a broadcast network's reach. But the combined audiences of cable networks, both pay and advertiser-supported, surpassed the broadcasters' totals.

Although a single broadcast network might still deliver 15 or 20 million viewing households each evening, the 100 or more cable networks often collectively attract more people than are aggregated by the combined five broadcast networks. Broadcast shows still routinely dominate the Nielsen ratings list of the Top 100 weekly programs but advertising dollars are being siphoned away. Cable systems have the added advantage of not only enjoying revenues from national and local advertising sales but also the constant stream of monthly subscriber fees. This dual revenue stream, compared to network and local television stations' single stream from advertising alone, made broadcasters start to consider a new strategy. Broadcasters no longer were willing to give their local cable systems access to their broadcast signals for free. They wanted a share of cable's monthly subscription fees.

As discussed in Chapter 4, the primary weapon in this conflict were the re-transmission consent regulations that could force cable systems either to pay local broadcasters for their signals or else not carry the local broadcast signal on their systems. In a fight played out in 2005, Nexstar Broadcasting pulled its signal from cable systems in Texas, Louisiana and Missouri until the cable companies there agreed to pay for the local Nexstar stations. Although the cable systems held out for months, Nexstar eventually signed re-transmission consent agreements with the cable operators that resulted in payments to the broadcasting company estimated at $10 million annually. CBS predicted that it would use a similar tactic in larger markets as its cable carriage agreements expired. This new source of revenue for television stations could be a substantial windfall – often amounting to millions per year.

New Technologies Change the Business

Just as television transformed the radio medium and cable changed television, a number of even newer technologies have and will continue to alter pre-existing media enterprises. A decade ago, Blockbuster and Hollywood Video provided consumers an alternative to television by renting first videotape, and then DVD copies of movies and previously-released television programming. Then Netflix, offering overnight distribution of movies via the postal service, eliminated the drive to the video store and gave consumers still more alternatives to broadcast or cable network programming. Today, *digital video recorders* (DVRs) coupled with fiber optics and massive server capacity threaten the very existence of brick-and-mortar video stores as well as the Netflix model.

With the launch of the first small-dish satellites over a decade ago, cable and traditional broadcast interests found themselves facing still other competitors in the form of DirecTV and Dish Network. These rival satellite (DBS) providers carry the same cable networks as do local cable systems – and compete directly with these systems in terms of price and picture quality. Based on successful international satellite ventures, primarily those in Europe and Asia owned by FOX's Rupert Murdoch, DirecTV, (now also owned by Murdoch), and Dish Network have significantly undercut the cable industry's dominance.[5] And the marketplace is growing more crowded by the day. Telephone companies, re-wiring the nation with their own fiber optic plants, can deliver video as well as voice and data. The result is a cable vs. broadcast vs. satellite vs. telco competitive environment where many business plans constitute major technology and consumer acceptance gambles.

Given streaming video on the Internet, portable playback devices for both audio and video, wireless technologies, satellites and other constantly mutating media, the mass audience is only a lingering memory. Media are all personal now. Technology is changing how we use electronic media, how and where we access these media, and how we are willing to pay for their content. One analogy that many media observers use to describe this technological shift is the change from a "push" model of media marketing to a "pull" model of media consumption.

In a "push" model, networks, movie studios and local stations push content down a distribution pipeline to consumers. Consumers simply tune in, buy a ticket, or change a channel to select an option pre-packaged for them by the distributing company. But today's quest to use media content how, when and where we want it has changed the paradigm.

Consumers are no longer willing to wait until dinnertime to watch the evening news, or to have it already packaged within the confines of a 30-minute program. When someone desires to know what is happening in the world or what team won the game, they want that information now. This is changing the electronic media's entire business model. No longer is the mass audience the target. The "pull model" game is now all about personal packaging and consumption.

Videotape and video recorders originally gave consumers the power to time-shift their television viewing. They could record their favorite programs for playback at a more convenient time. Newer devices, such as TiVo and similar digital video recorders (DVRs), make it easier for consumers to shift viewing times and avoid commercial interruptions. Larger hard drives in these devices or DVRs based on server technology at the cable system's head-end both allow thousands of hours of movies and television programs to be recalled, played, paused and viewed whenever and wherever the consumer chooses.

In addition to *time-shifted viewing*, consumers now expect *place-shifted viewing*. Rather than watching television in the family room as a collective experience, some consumers expect to individually access their favorite television programs on their personal computers in a wireless environment. Whether they are students killing time between classes, or people at work or in transit, consumers increasingly will use television very differently than in the past. Consumers also expect media to be everywhere. The thought of living without a broadband connection, not having a 100-plus-channel cable or satellite system subscription, or being deprived of portable viewing and audio options is simply unthinkable to these new media consumers.

On the radio side, XM and Sirius, the two national satellite radio companies, have created a pay radio model to compete with traditional radio broadcasters. The FCC licensed these two companies in the mid-1990s to build and operate multi-channel, pay radio services delivered by satellite. Early in the new century, both XM and Sirius were delivering one hundred or more channels at a monthly charge of $12.95. The automobile companies have been proponents of the firms by equipping their new vehicles with one or the other of these pay satellite radio offerings,

Such *"pay radio"* services, with a much broader offering of commercial-free music formats, major league sports from every team in the nation and edgier content from a variety of former radio personalities on their talk format channels, are the darlings of Wall Street and command larger market values than the top 10 radio companies. Although both XM and Sirius currently lose money, investors have snapped up their stock and driven up the price of their shares to astronomical levels. XM and Sirius are betting that consumers will pay for audio programming much like they pay for television and movies.

Traditional media companies now are learning to re-purpose their programming so that a television show can be streamed on the Internet following its broadcast on the main television channel. Similarly, radio stations are offering podcasts or mini-programs to be downloaded into personal iTunes players. Through such mechanisms, these older media outlets are trying to reach consumers wherever they may be. Thus, Gannett, a major newspaper and television station owner, operates *Captivate*, a business that places television displays in high-rise office elevators. Gannett's goal is to capture people for a few moments at a time with entertainment and information snippets, along with commercial messages; all transmitted to them in the inescapable confines of an elevator compartment.

Meanwhile, Apple's iTunes created a demand for MP3 audio that could be downloaded and played on a micro-sized device. Then Apple began downloading television programs and video podcasts as well as audio files to allow consumers to make their television viewing experience personal, portable and on demand. Consumers appeared willing to pay for this convenience of time and place-shifted viewing as evidenced by Apple's report of 14 million iPods sold for Christmas, 2005.

Network executives are experimenting with how much to charge for personal viewing, whether the original show or expanded versions work best for personal consumption and if alternative story lines and endings might interest the public. They are also looking for new revenue sources beyond traditional commercials carried within the program. Product placement, for instance, is a booming phenomenon. Given that so many consumers are skipping commercials as they watch recorded programs on their DVRs, advertisers are placing their products directly into the story line. Product displays and mentions are up significantly in recent years as advertisers fight to break through the commercial clutter and ad skipping behavior afflicting traditional media.

Vertical Integration as a Survival Strategy

One critical element in the development of media businesses over the past 20 years has been the vertical integration of these companies. *Vertical integration* is the common ownership of the production, distribution and marketing functions. This means that one company may produce the program or movie that is then distributed on its own television or cable network. It also sells the advertising for the television program and may control the sale of copies of that program or movie into domestic and international syndication as well as directly to consumers. This vertical integration sometimes extends to ownership of radio stations and magazines capable of promoting programming – or to book publishing companies that may provide a rich source of original material for electronic media projects.

Most television shows are produced in Hollywood by either the television production divisions of the major motion picture studios or by independent producers under contract with the television networks. As we discussed in Chapter 3, after initially refusing to sell their movies to the television industry, the major motion picture studios began churning out thousands of hours of programs for the television networks in the mid-1950's. Over the past 20 years, movie studios have themselves built or purchased television and cable networks – or been purchased by network interests. In many cases, these companies now own a motion picture studio, several television and/or cable networks, cable systems, a radio station group, book and magazine publishing ventures, and a variety of other media businesses.

Today, there are seven companies with strong international business platforms that dominate entertainment production and programming. These companies are Time Warner, NBC Universal, Viacom, CBS Corp., News Corporation, Sony and Disney.[6] Sony, as a Japanese company, is not allowed to own US broadcast stations and is therefore at a competitive disadvantage compared to the other multinational entertainment companies that own either broadcast stations and networks or cable systems and networks. On the other hand, Sony possesses a strong hardware technology division that the other six companies lack.

Vertical integration of the entertainment and media business provides two advantages to these companies compared to independent production companies or media interests lacking any major production element. First, there is a strong ability to cross-promote the brands and programming of one medium on other media owned by the same conglomerate. Movies can be promoted in television infotainment programs and pushed in radio and magazine interviews. Television shows, records and other assets owned by a motion picture company can similarly be promoted in pre-feature advertisements at the movie house.

This vertical integration of media also accrues a second advantage by creating a seamless pipeline for distribution of productions. Disney owns the ABC network, News Corporation owns the FOX network, CBS Corp. owns CBS and half of The CW network, while NBC Universal owns NBC plus the Spanish-language Telemundo network. Each media company has an economic incentive to produce in-house programs or to purchase programs from the parent Hollywood studio for broadcast on the network because this keeps all revenues "within the family."

Table 5.1 charts the holdings of the seven major media companies, which all operate on a global basis. It demonstrates the pervasiveness of their influence on the entertainment and news programming that millions of people consume daily. FOX has subsequently added MyNetworkTV to its broadcast TV roster.

A Smaller Slice of a Larger Pie

Broadcasters have seen their share of listening and viewing steadily erode over the past two decades. The advent of basic and then pay cable, direct-to-home satellites, video rental stores and now broadband video on demand, has dramatically altered the landscape for traditional networks and local stations. The mass audience is disappearing and personal media have arrived. Broadcast television is still capable of delivering gigantic audiences for special event programming and remains an efficient delivery mechanism for many advertisers. The broadcast networks remain dominant compared to any one cable network's program ratings. However, as previously noted, the vast number of cable networks, taken together, today reach more consumers than do their five broadcast counterparts

Because of population growth and the steadily-increasing aggregate amount of television viewing by the public, individual media outlets may see their share of the total viewing audience becoming smaller while the overall media consumption pie grows every year. More consumers, coupled with longer viewing habits, result in a larger pie. Broadcasters, who consumed almost the entire pie 30 years ago, command a smaller piece of a larger pie today. Advertising rates continue to rise and network sell-out levels continue to be solid.[7] For many advertisers, broadcast network television remains the single most effective advertising medium available. Broadcasting will, no doubt, continue to be a very attractive business for many years. Particularly if new re-transmission consent agreements are struck with local cable operators to carry their stations, broadcasters will continue to be prime local video and data content providers.

The concern for broadcasters is that technological shifts are permanently changing some basic viewing and listening behaviors. This is

Table 5.1 The scope and diversity of the seven major media companies

	Time Warner	NBC Universal	VIACOM	CBS Corporation	News Corporation	Disney	Sony Corp
TV Networks	The CW (50%)	NBC		CBS, The CW (50%)	FOX	ABC	
TV Group	Time Warner Inc	NBC/GE	VIACOM		Fox TV	ABC/Disney	
Radio Networks	CNN Radio	NBC Radio		CBS Radio	Fox Sports Radio, Fox News Radio	Radio Disney, ESPN Radio, ABC News Radio	
Radio Group						ABC/Disney	
Cable Networks	Cartoon Network, CNN, TBS, TNT, Turner Classic Movies, Court TV, Boomerang, TNT HD, Turner South	Telemundo, Trio, Bravo, CNBC, CNBC World, MSNBC, mun2, Sci-Fi, Universal HD, USA, NBC Sports, ShopNBC, Universal HD	MTV, MTV2, VH1, Nickelodeon, Nick at Nite, Comedy Central, SpikeTV, BET, Logo, The N, TV Land, CMT, MTVu, Noggin, Flix, Sundance Channel	CSTV	Fox Movie, Fox News, Fox Sports, Fox College Sports, Fox Reality, FUEL TV, FX, National Geographic, FOX Kids, SPEED, Stats, Inc.	ESPN, Disney Channel, ABC Family, Toon Disney, SoapNet, Lifetime TV, E!, A&E, History Channel, ABC Sports, Classic Sports Network	Game Show Network (GSN)
Pay TV	HBO, Cinemax	Pay-Per-View & On Demand		Showtime	Fox Pay-Per-View		
Motion Picture Production	New Line Cinema, Picturehouse, Warner Bros. Pictures, Warner Independent, Castle Rock Entertainment, Fine Line	Universal Pictures, Focus Features, Rogue Pictures	Paramount Pictures, Paramount Classics, DreamWorks SKG		20th Century Fox, Fox Studios LA, Fox Searchlight Pictures, Blue Sky Studios	Walt Disney Pictures, Touchstone Pictures, Caravan Pictures, Miramax, Dimension, Walt Disney Feature Animation	Sony Pictures, Entertainment, Columbia TriStar Motion Picture Group, Sony Classics, Sony Pictures Digital

TV Production	New Line TV, Warner Bros. TV, Telepictures TV	NBC Entertainment, NBC Universal TV Studio		Paramount TV, King World	20th Century Fox TV, Fox TV Studios	Touchstone TV, Buena Vista TV	Sony Pictures TV Group, Columbia TriStar
Music Production	New Line Music			Famous Music	Festival Mushroom Records	Walt Disney Records, Buena Vista Records, Hollywood Records, Lyric Street Records, Mammoth Records	Sony BMG Music Entertainment, Sony ATV Music Publishing
Distribution	New Line Distribution, Warner Home Video	NBC Universal TV Distribution, Universal Studios Home Entertainment		Paramount Home Entertainment	20th Century Fox Home Entertainment	Buena Vista Home Entertainment, Touchstone Distribution	Sony Pictures Home Entertainment, Columbia Tristar Home Entertainment, Sony Pictures TV
Print Publishing	Time Inc; Time Warner Book Group, Warner Bros. Publishing, American Family Publishers		MTV Books	Simon & Schuster	TV Guide, The Weekly Standard, SmartSource, New York Post, HarperMorrow, ReganBooks, Zondervan	Hyperion Books, Disney Educational Productions, Disney Press	
Cable MSO/ Digital Broadcast/ Satellite Co.	Time Warner Cable Inc, Road Runner, Road Runner-Business				DIRECTV		
Interactive Services	AOL, The Knot, WB Internet, Music Choice (joint), Digital City, CompuServe, MapQuest, ICQ, Moviefone, Netscape, Movielink (joint), McAfee VirusScan	NBC Internet, ShopNBC, Movielink (joint)	Movietickets.com, Movielink (joint), MTV Networks Online, Viacom Interactive, GameTrailers.com, IFILM.com, Neopets	CBS Digital Media	MySpace.com, Fox Interactive, Scout Media, IGN Entertainment	Disney Online, ABC Internet Group, NASCAR.com, NBA.com, Toysmart.com, Family.com, ESPN Internet Group, Go Network, Wall of Sound, Skillgames	Sony Internet, Movielink (joint), Sony Connect, Music Choice (joint), 989 Sports, Sony Online Entertainment, SoapCity

Table 5.1 *Continued*

	Time Warner	*NBC Universal*	*VIACOM*	*CBS Corporation*	*News Corporation*	*Disney*	*Sony Corp*
Other	Hanna-Barbera Cartoons, Time Warner Telecom, Content Guard, Warner Bros. Theatre, Atlanta Braves, Time Entertainment	Universal Parks & Resorts, NBC Olympics	National Amusements Inc	CBS Outdoor, Paramount Parks	NDS, National Rugby League, Staples Center (joint), LA Lakers (joint), LA Kings (joint)	Disney Parks & Resorts, ESPN Zone, Disney Institute, Buena Vista Theatrical Productions, Anaheim Sports, MGM Studios	Sony Pictures Digital, Sony Pictures Mobile, Sony Financial Services, Play Station, Metreon, SpiN, Sony Online Entertainment
International	AOL Europe, HBO Intl, WBTV Latin America, El Latin America, Time Inc, Cartoon Network Intl, CNN Intl, TCM Europe, TNT Latin America, WB Intl Cinemas, Time Warner Book Group UK, Compuserve Intl, Warner Bros. Pictures Intl	Universal Studios Networks Intl, CNBC (Europe, Asia Pacific), Telemundo Intl, NBC Europe, NBC Universal Entertainment Europe, Universal Studios Japan, Universal Mediterranea	TDI Worldwide, MTV Networks Intl	Simon & Schuster (UK, Australia) Viacom Canada, CBS Paramount Intl TV	20th Century Fox Intl, Fox Studios Intl, STAR, Fox Sports, FOXTEL, UK & Australia newspapers, Australia magazines, Harper Collins (Canada, Australia, UK), Sky Radio, Radio Veronica, News Outdoor, Classic FM, SKY, NDS, Broadsystem, News Interactive, News Optimus	Buena Vista Home Entertainment Intl, Buena Vista TV Intl, ESPN (Brazil, Asia), Disney Publishing Intl, Disney Parks & Resorts, Disney Channel (UK, Taiwan, Australia, Malaysia, France, Middle East, Italy, Spain), Net STAR, Sportsvision of Australia	Sony Pictures Television Intl, Sony Corp, Sony Ericsson Mobile Comm, Sony Europe, Sony of Canada, Animax Japan, Sony UK, Sony Asia Pacific, Sony Japan, Sony Latin America, AXN, SET (Latin America, India)

perhaps clearest in the world of radio and music. Today, an entire generation of people are turning away from radio as a primary source of music. They are instead relying on iPods and downloads of music to their computers and MP3 players in order to satisfy their music appetite. Couple these technologies with the birth of satellite radio, stir in the exodus of many younger-appeal talk personalities to major satellite radio providers, and radio station operators are rightly concerned. Efforts to roll out HD multi-channel digital radio may already be too little and too late for radio station owners. Technology has shifted the way an entire generation uses media for music consumption.

On the television side, broadband video-on-demand and the Internet pose behavior-shifting challenges to broadcast as well as cable television. As flat-screen and high definition television become married to home computers, it will be increasingly easy for consumers to write and receive e-mails, conduct web business or shopping and download movies and other video programming on their television screens as opposed to watching time-specific television or cable channels.

Radio and television stations remain healthy businesses with attractive profit margins and the ability to benefit from three key changes in the media landscape. The *first change* involves new regulations which may allow television and radio station groups to grow even larger and to benefit from economies of scale. The regulatory approval to operate two or more television stations, as well as a newspaper, several radio stations, a billboard company and perhaps even a cable system all in the same market would be very beneficial from a financial standpoint. But such concentration raises serious consumer and advertiser concerns about the free marketplace of ideas as well as the pricing power a single company might hold.

A *second change* in the media landscape may allow stations to participate in a share of cable television revenues through re-transmission consent deals with the cable industry. Rather than simply giving away their signals to cable companies for free, broadcasters have begun to demand payment for their signals via new re-transmission consent negotiations. The Nextar fight in 2005, and CBS's vow to extract re-transmission consent payments from cable, are the initial battlegrounds for this regulatory and economic fight. Broadcasters have been slow to understand that they should not simply give away their signals to another company that charges consumers for this content.

The *third change* in the media landscape is occurring as television broadcasters learn to: (1) use the Internet to drive audiences to and from their stations; (2) sell their programs "on demand;" and (3) expand into original web productions that can aggregate viewers and generate subscription dollars. Stations in a number of markets now routinely turn informational programming into longer-form video presentations for streaming on the Internet while charging sponsors for this service.

For example, after producing a multi-part series on laser eye surgery for their on-air newscast, and following a patient through the entire process, television stations can then expand this video into a 30- or 60-minute program and stream it on the station's website. DVD copies of the longer program can also be provided for a fee to the physicians who cooperated in the preparation of this show. The physicians also can sponsor the webcast and purchase regular advertising time on the television station.[8]

Television outlets are realizing that the Web can be a significant advertising vehicle that complements conventional television advertising. Whether they are walking patients through laser eye surgery, presenting videos by attorneys and physicians on living wills, advertising automobiles for sale via digital still pictures or streaming video, presenting interior design ideas, selling books and videos, or offering travel planning for family vacations, broadcasters can exploit the Internet as a powerful revenue tool. Already, television station operators in medium to large markets are generating several million dollars in annual revenues from the web.

Today, traditional broadcasters are experiencing fundamental changes in their 80-year-old business. New media present both new challenges and new opportunities. Some may long for the days of only three major networks, little competition and large, predictable cash flow margins. Modern media operators, however, recognize that consumers are driving their businesses. More choices lead to more innovation and a quicker response to changing markets. Consumers are demanding media on their own terms but are starting to realize that paying for content, rather than receiving it free with embedded advertising, is also part of the landscape. Time-shifted viewing, the choice of watching video programming either with embedded commercials or on a commercial-free fee basis, as well as the ability to personally archive thousands of programs, are all realities for today's consumers. These are also the realities for media operators. It is a brave, new world for these operators that demands careful, yet courageous financial planning.

Chapter Rewind

From the beginnings of radio in the 1930s to the birth of television in the early 1950s, broadcasting has always been an extremely profitable business. The demand for broadcast licenses, particularly television licenses, was enormous. As a business featuring the dual advantages of limited entry due to federal licensing and a fixed cost of operation capable of generating enormous profits, broadcasting became a very attractive enterprise.

Local television stations depended on the networks for much of their programming throughout the day but also found great profit centers in local news and syndicated programming. With the advent of the national cable networks, the broadcast networks began to lose viewers and market share. Yet, advertisers continued to look to broadcast television as the primary national mass medium. New technologies presented challenges but also opportunities, such as downloading television programs to iPods and digital video recorders for playback at the consumer's convenience. Local television stations, however, continued to generate substantial audiences and revenues through strong marketing, expanded local newscasts and attractive syndicated programming. On the national level, vertical integration escalated with many of the television networks becoming part of larger entertainment companies owning movie studios, cable networks, radio groups and other media.

Broadcasting and cable fought many regulatory battles over the past decades. Retransmission consent, through which broadcasters look to be paid for cable's use of their signals, promises to create a seismic shift in the economic relationship between these parties.

SELF-INTERROGATION

1 Explain the relationship between networks and affiliates.
2 Why are the broadcast networks' O & O divisions more profitable than the networks themselves?

3   Explain the difference between "push" and "pull" concepts of media usage.
4   What developments are diminishing the historic power of the television networks?
5   How are the television networks facing the challenges of new media?
6   What have been the three tactics for success as a local television affiliate?
7   How exactly did cable split the mass audience?
8   How and why did the financial viability of network news change?
9   Select three new media technologies and explain how they both splinter the mass audience and also generate new revenues for themselves.
10 What is "vertical integration" and how does it operate in media companies?
11 What are three regulatory or business changes that local television stations face as they move into a world of personalized media and as they attempt to stay both relevant and profitable?

NOTES

1   NBC's "up front" sales for the Fall, 2005 season were reportedly off by at least $1 billion based primarily on weak ratings and consequent lack of advertiser demand for their series programming.
2   It was estimated that each of the three original television networks was spending at least $200 million on affiliate compensation payments by 2000. The FOX network, with fewer stations and a more shaky financial condition, had never paid much compensation and was well below this figure.
3   By 2000, FOX moved quickly to consolidate these regional pay cable sports networks and to brand them under the FOX Sports logo.
4   Spike TV was originally The Nashville Network, a country music and entertainment service, and later was The National Network before Viacom focused it as a male-oriented cable offering and dropped any country-music connection.
5   Nielsen reported in late 2005 that several markets, including Springfield, (Missouri), Missoula, (Montana) and Butte, (Montana) actually had more satellite dish subscribers than traditional cable subscribers.
6   At the end of 2005, Viacom split into two companies – Viacom and CBS Corp. Sumner Redstone, controlling shareholder of Viacom, said that the move was made to capture the stock market's under-valuation of its cable television network assets, such as MTV and Nickelodeon. Shareholders of Viacom were given stock in both companies. The initial pricing of the individual stocks were higher for Viacom (the new media company) than for CBS Corp. (the old media company). The new Viacom was predicted

to have more growth in future years while CBS was expected to have very stable cash flows and bedrock value. CBS has already begun to re-enter the cable network business, however, by purchasing a small college sports network. However, CBS has said that it would stay out of music and children's cable where Viacom remains dominant.

7   Newspapers have experienced the same phenomenon. Circulation levels have steadily dropped over the past 10 years. Yet, advertising rates have generally at least doubled during the same period.

8   Some examples of this are KLAS-TV (Las Vegas); WAVE-TV (Louisville); and KETV-TV (Omaha). Each of these stations did a laser eye surgery story followed by a streaming video webcast. In some cases, the physicians featured in the program paid the station a six-figure advertising fee.

PART II

Challenges

CHAPTER 6

Technological Challenges

Steven D. Anderson

The world of electronic media was a much simpler place when dominated by traditional radio and television stations. Changes caused by the Internet, video games, MP3 music players, PDAs and cell phones have been as profound as the impact of earlier inventions such as the telegraph, telephone, and wireless.

Stimulated by the explosive popularity of products like digital television sets, digital video recorders and small portable gadgets including iPods, the average American family has come to own 25 consumer electronics devices.[1] These technologies could be viewed as threats or opportunities for traditional broadcasters, but it is clear that formerly passive listeners and viewers are today demanding an active role in their interactions with the media. Another challenge involves the proliferation of alternative delivery systems such as satellites and the Internet. Consumers today can choose between over-the-air broadcasting, cable television, direct broadcast satellite (DBS) and a range of other options including both wired and wireless broadband Internet. There is tremendous marketplace competition for the delivery of multi-channel audio and video programming.

In the following sections, we'll look at current technological challenges in three major areas: (1) changes to traditional media; (2) device convergence; and (3) Internet and broadband advancements.

Changes to Traditional Media

Satellite radio

As the century turned, radio technology had remained largely unchanged for many decades. But the 2001 arrival of *satellite radio* broke new ground with multiple high-quality digital channels beamed from space directly to consumers. Satellite radio is described in terms of the traditional sound medium and comes into cars and homes much like the terrestrial radio we've known for more than 80 years. A distinguishing feature is satellite radio's ability to cover a nationwide audience. It allows users to listen to one station continuously while driving great distances; freeing radio from the limitations of reception from a local tower. Users pay a monthly subscription fee for the service and can receive more than 100 channels of usually commercial-free audio programming including news, sports, talk and a diverse selection of music formats.

Two companies initiated satellite radio. XM Satellite Radio was the first, launching in 2001, followed the next year by Sirius Satellite Radio. Both operate digital broadcast centers in the United States where separate recording studios create the individual channels. The channels are then compressed to fit within the 12.5 megahertz of spectrum allocated by the FCC and uplinked to satellites, along with a short description of the artist and song that can be displayed on the listener's radio. Although the signal doesn't depend on land-based transmitting towers as do conventional stations, the satellite transmissions are nonetheless augmented by an extensive repeater network on the ground to assure full coverage to subscribing cars and homes. This is especially useful in urban areas where line-of-sight satellite reception can be interrupted by buildings, overpasses and other obstacles. Radio receivers are equipped to receive the signals directly from the satellite, but if the signal becomes blocked, the receiver automatically switches over to reception from one of the repeaters. Several seconds of audio is stored, or buffered, in the receiver so the programming can continue even if no signal is available for a short period of time.

XM Satellite Radio, based in Washington, DC, uses two satellites named "Rock" and "Roll". They are in geosynchronous orbits over the equator, rotating at the same speed as the earth and therefore continuously "parked" over the US. But Rock and Roll's low position in the sky relative to the United States requires XM to use an extensive network of ground-based repeaters in order to provide uniform signal coverage. Sirius Satellite Radio, based in New York City, does not use geosynchronous satellites. Instead, three satellites are equally spaced in elliptical orbits that take 24 hours to complete one rotation around the earth.[2] These orbits take the satellites over the United States sequentially, with at least one overhead at any given time. Therefore, Sirius requires fewer repeaters to maintain an optimal coverage pattern. The digital audio signals of both companies are encrypted, so that their service will only be available to paid subscribers who have special receivers to decode the signal. XM also offers handheld portable devices for those wanting to take their music with them. Portable XM receivers can tune in all of the XM channels and even record more than one station at a time, holding as much as 50 hours of recorded content in built-in Flash memory.

Satellite radio providers are gearing up for a major push to make their services more interactive. XM is introducing a feature that allows drivers to switch stations using voice commands.[3] The company also plans to offer extras such as real-time traffic information, advance weather warnings and a search feature to find available parking. Users will one day be able to receive and record video feeds as well. According to Forrester Research 20.1 million households will regularly listen to satellite radio by 2010.[4]

HD radio

Providers of satellite radio have grabbed all the attention on the radio side, but land-based (terrestrial) local radio stations have found a way to get into the digital ballgame with *high-definition (HD) radio*. High-definition radio lets stations broadcast high-quality (near CD level) audio, along with textual information such as song title and artist name. HD radio also provides a range of advanced features, like on-demand interactive and wireless data services including stock quotes, weather information and traffic alerts. Similar to what digital technology is doing for television, HD further allows local radio stations to engage in multicasting – essentially letting one outlet broadcast multiple programming streams within the same amount of spectrum space it previously occupied. Thanks to digital compression technology, up to eight program streams can be squeezed into a single channel assignment. This makes a local radio station a multiple program provider much like a satellite radio company – though with far fewer program choices.[5]

Through such technology, AM radio stations, which use less bandwidth than their FM counterparts, will be able to improve the fidelity of their signals to rival the quality of conventional FM stations. This could assist AM to become more competitive with FM in conveying music programming. The cost for a station to convert to HD radio is far less than the cost associated with a television station converting to digital television. Also unlike digital television, radio stations get to choose whether they want to upgrade to HD radio, or stay with their current analog broadcasts. Of course, new radio receivers are needed to pick up the HD radio signals. Currently, the price of HD radio receivers for cars and homes is high. However, unlike satellite radio, listeners to HD stations listen for free. As HD radio receivers are built into more cars and personal listening devices such as MP3 players, the battle between free land-based local radio stations and subscription-based satellite radio providers will likely intensify.[6]

Digital television

While the regulatory and economic challenges of digital television are still being worked out, the technology behind DTV is largely in place. The standard originally adopted in 1996 has provided broadcasters with a set of technological tools to create stunning images and sound. For television stations, the current challenge is finding money to buy DTV cameras, tape machines, storage units and a wide range of studio and field production equipment. While the networks have determined their own implementation of the standards, local stations can choose to purchase equipment capable of 480 lines, 720 lines or 1,080 lines. The more lines of resolution, the more the equipment will cost. Many small market television stations are purchasing standard-definition digital equipment capable of only 480 lines. Larger market stations may choose to buy high-definition equipment at 720 lines or 1,080 lines. Whatever decision a station makes, the signal can always be *up-converted* or *down-converted* and transmitted at a resolution different from the original *native resolution*. For example, a program produced on native 480 line equipment can be up-converted and broadcast at either 720 lines or 1,080 lines. Of course, the up-converted signal won't look as good as if it had been produced originally at a higher resolution.

For the consumer, choosing between CRT (cathode ray tube), plasma, LCD (liquid crystal display), DLP (digital light processing) or SED (surface-conduction electron–emitter display) sets is just the beginning. All may be digital displays, but some are only capable of showing images at 480 lines. Even if a program was produced and transmitted at 1,080 lines, those 480-line sets won't be able to display them. Also, owning a digital television set isn't enough to guarantee being able to watch DTV signals on it. Determining what additional equipment might be required depends on two factors: (1) how the signals will be delivered (over-the-air, cable or satellite); and (2) the type of digital television set purchased (integrated, HD-ready or HD-compatible).

Those who expect to watch DTV directly from over-the-air broadcasters will need a DTV antenna and an *integrated digital television set,* which includes a built-in DTV tuner. (The FCC has ruled that every television set manufactured after March 1, 2007, must have a built-in DTV tuner). Television sets rated as *HD-ready* or *HD-compatible* will require a separate DTV tuner. Those who expect to receive DTV signals from cable will need a digital cable box or a cable card, while those receiving DTV channels from satellite will need a DTV satellite box. These boxes are available to buy or rent from the service provider.

By February 17, 2009, broadcasters will be required to give back their analog NTSC channel. (UHF channels above channel 51 will be sold to mobile communications companies for emergency "first-responder" and mobile phone services). At that time, people who wish to view over-the-air DTV signals on their NTSC sets will be able to purchase a low-cost converter box, but won't receive the benefits of clearer, higher quality DTV signals.

For its part, the cable television industry has been slow to deliver digital television (DTV) signals on their systems. While they've been providing a service called *digital cable* for a number of years, digital cable is nothing more than delivery of analog NTSC channels via digital signals. Cable operators have been hesitant to carry the new DTV signals from local broadcasters because those signals essentially duplicate the NTSC programming already carried on their systems. Carrying both the broadcasters' DTV and NTSC signals would limit the number of different programs cable systems could offer to subscribers. In 2009, when analog broadcasts end, cable operators will have more incentive to convert, but delivering DTV signals on cable still will come at a high cost. Cable operators need to replace millions of old analog set-top boxes with digital boxes before they can turn off their analog feeds.

One contentious issue between broadcasters and cable companies involves the potential for *material degradation*. After the 2009 analog cutoff date, cable companies could convert a broadcaster's digital signal to an analog signal for carriage on their systems. A broadcaster's signal could be further compromised if cable companies choose to use digital compression technology in an attempt to shoehorn even more channels onto their systems. Compression can squeeze as many as 10 channels of high-definition television into the same 6 megahertz of space currently carrying just one NTSC television channel. Thus, cable companies could theoretically accommodate a thousand or more channels of high-definition television on their systems. Of course, compressing video comes at some sacrifice. Too much compression results in *compression artifacts* – a visible corruption of the image sometimes referred to as *pixelization*. Broadcasters would prefer that cable companies pass through their signals in digital form and that they transmit all of the digital bits to prevent material degradation.

Direct broadcast satellites

In the past few years, *direct broadcast satellite* (DBS) technology has developed the capability to provide local signals to some markets. Instead of sending thousands of local TV stations indiscriminately to the entire country, a *spot beam* works rather like a flashlight to focus the signal from a satellite to a spot in the country between 100 and 200 miles wide. This allows the DBS operator to rebroadcast on the same frequencies in different locales, thus making more efficient use of limited spectrum. For example, they might send the local Denver stations to Denver homes on one set of frequencies and use those same frequencies to send consumers in Chicago and New York their local stations. With digital compression and greater satellite capacity, the DBS industry believes it can soon provide approximately 1,500 local channels. DIRECTV claims to already offer local channels to over 90 percent of the US households. These local station packages enable DBS companies to more effectively compete with cable systems.

Digital broadcast satellite providers DIRECTV and Dish Network are currently offering a limited number of high-definition digital television signals as well. Via special HD satellite receivers, both DIRECTV HD and Dish Network HD offer high definition versions of network programming, sports, movies and original features not found on other outlets. In some television markets, local channels are also available in HD via dedicated receivers and dishes.

Device Convergence

New media technologies provide more than simply an improved listening or viewing experience. They are also drastically changing the way users consume media, turning what was a passive experience into a media-rich world where customers control content. Media users today can decide for themselves what they consume, when they consume it and on what kinds of devices. Media can be delivered via terrestrial broadcast, cable, satellite, phone lines, wireless or even the mail (i.e. NetFlix). Electronic content can be consumed on an array of devices – TiVo, iPods, cell phones, computers, television sets and even game consoles like Sony's PlayStation. Chosen content can be viewed live or delayed to fit people's individual schedules. And it can be customized or personalized to suit their preferences. Those who produce programming must understand the more active role media users play and reach them via the avenues they prefer. These changes are about technology, but more importantly they reflect the significance of content created specifically for new delivery devices and viewing patterns.

The word *convergence* is used to describe the integration of personal communication, computing and consumer electronics on a single device or delivery system. Technologically, *device convergence* is affecting every aspect of traditional media including advertising, news and programming.

The functions found on separate appliances such as cell phones, digital cameras and PDAs (personal digital assistants) are today being integrated in a single *hybrid device*. One can talk, schedule appointments, look up addresses, view Web pages, send and receive text messages, listen to music, acquire GPS (global positioning system) coordinates and take photos on multi-functional (converged) mobile instruments. Several companies have gadgets that combine a music player, a digital camera and a mobile phone.[7] Others offer mobile phones that carry digital music, photos and video, along with FM radio signals.[8] Many of these devices can exchange information with each other, or with computers and printers, via *Bluetooth*, a short-range wireless network connection. Features are expanding as mobile phone companies improve their wireless networks to include broadband capability. For example, a newer broadband wireless protocol called *EV-DO* (Evolution Data Optimized) provides enough data for live video feeds on a mobile phone.[9]

Even the ubiquitous game console is getting into the device-convergence act. Both Microsoft and Sony produce converged consoles capable of playing games, listening to music, watching movies and displaying photos on network-connected devices. Microsoft's *Xbox 360* broke new ground when it provided expanded capabilities for Web services, email and search technology.[10] Portable game consoles feature games, music, movies and photo capabilities and now provide wireless WiFi connections. The devices offer a level of mobility previously associated only with cell phones and MP3 music players.[11] Hardcore techies soon extended the capabilities of the Sony PlayStation Portable by building add-on features Sony never intended such as Web browsing, chat, access to news feeds, and even a way to transfer TV shows recorded on TiVo.[12]

Televisions and computers are combining functionality. For a number of years, television sets have been able to view Web pages and offer interactive features such as *video-on-demand* (VOD). Likewise, computers have been able to receive television content (episodic programs, movies, commercials) via audio and video *streaming*. These capabilities are being expanded as software and hardware manufacturers eliminate the need for separate cable boxes to receive digital video signals for television set display. In this scenario, a cable line plugs directly into a PC, turning the computer into an HD- capable cable box and allowing it to distribute content to multiple screens including the computer, television, game consoles and other "boxes" throughout the home.[13] One can use a PC to receive the signal, store video and manage where it will be shown; combining video, data and even voice on one converged computer-centered media utensil.

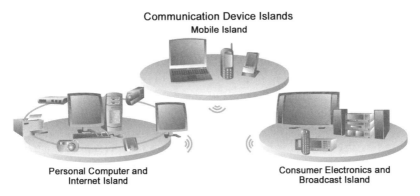

Communication Device Islands
Mobile Island

Personal Computer and
Internet Island

Consumer Electronics and
Broadcast Island

Figure 6.1   Segregation of consumer products. Consumer products tend to be segregated onto three islands. Getting these products to interoperate and communicate with each other across the islands is a key convergence challenge. Source: Courtesy of the Digital Living Network Alliance, "Overview & Vision White Paper," copyright 2004. All rights reserved.

Computer chip makers are creating chips to further enhance convergence. Intel's Viiv (rhymes with "five") technology is designed to make computers more compatible with television-based *digital video recorders* such as TiVo, allowing easy movement of files between PCs and external devices, all from the comfort of a remote control.[14] The technology will allow a PC to operate more like a TV with an "instant on" feature that essentially awakens a PC from a sleep setting. Dozens of movie, music, gaming and photo-editing products can also be configured to work with the technology.

In order for these digital products to take hold in the home, consumer frustration over incompatible technologies will need to be overcome. Today, communication devices exist on three islands: the *PC island* of computers and peripherals; the *broadcast and consumer electronics island* of television sets, stereos, DVD players and DVRs; and the *mobile island* of laptops, mobile phones and PDAs[15] (see Figure 6.1). Getting these products to work together and communicate with each other (*interoperability*) can be daunting. Companies often create devices with proprietary standards, making them incompatible (digital camera memory cards are a good example). To solve the problem, industry groups set up the *Digital Living Network Alliance* (DLNA); an organization made up of companies representing all three islands. Through the

DLNA, firms such as Microsoft, Intel, Motorola, Nokia, Sony and Panasonic are working to develop standards and create products that are interoperable.

By 2006, traditional broadcasters were making their content available on more devices, striking deals with competitors such as cable providers and Internet companies to expand their methods of distribution and woo viewers they were in danger of losing to new technologies. For example, NBC Universal and CBS began making prime-time television shows available for download on the Web in a pay-per-episode scheme, without commercials. If a viewer misses a favorite program and fails to set up recording of the show in advance, it is available later for as little as 99 cents per episode.[16] Apple and ABC developed a pay-per-episode model that doesn't even require service from a cable company or satellite provider, instead allowing direct download of ABC prime-time shows for viewing on a computer or video iPod.[17] Apple and NBC Universal forged a similar agreement.[18] For viewers who didn't want to wait, DirecTV offered FOX's FX shows for download at $2.99 before they aired on broadcast television. Episodes of primetime shows on FX include additional scenes and materials and are available 48 hours before they air on the network broadcast.[19]

In some cases, content is being produced specifically for small-sized portable devices such as cell phones. Labeled as *mobile television*, both CBS and FOX have developed soap operas, or *mobisoaps*, for mobile phones using high-speed cellular connections. The productions have their own writers, cast and production teams. Episodes run from three to five minutes each, making them accessible by multitasking viewers engaged in other activities or just waiting in line.[20]

Internet giants Google and Yahoo! are joining forces with broadcast networks to bridge the gap between computers and television sets.[21] Google has opened an online video marketplace where content providers set the price, while sharing revenue with Google. The service provides a library of commercial free videos for purchase or rent. Content from broadcast networks, as well as sports programming, independent films and vintage episodes are available. Yahoo!'s Go TV service links computers with television sets, allowing a user to watch anything stored on a computer on the biggest screen in their home.

The fusion of network-connected computers, televisions and digital video recorders may make pay-per-episode schemes unnecessary. TiVo is offering a service ("TiVoToGo") that permits consumers to record their favorite shows and convert them to formats capable of being played on a computer, iPod or Sony PlayStation. With a little planning, viewers can record their favorite shows and bypass the pay-per-episode system.[22] To make the recording of shows even easier, TiVo and Yahoo! blended television and Web services, allowing users of Yahoo!'s TV page to click on a "record-to-TiVo" button directly from the television page listing.[23] Another device, called a *Slingbox*, enables people to view television on a computer or laptop anywhere in the home via a wireless connection. The gadget imports the over-the-air, cable or satellite television signal, compresses it into a computer-based video format and facilitates storing the video on a computer's hard drive or streaming it out live to computers within wireless range.[24] Devices like the Slingbox and Sony's *Location Free TV* even allow the user to *placeshift* by setting up the device at home and accessing it with an Internet-connected computer in another location. Anything you can watch at home, you can watch wherever you are around the world.

By 2006, the broadcast networks were already offering some programming without a pay-per-episode fee. NBC began making *NBC Nightly News* available free on the Web, immediately after it has been shown in all US time zones. NBC, ABC, and CBS had already provided selected video clips on their websites for a number of years at no charge.[25] Whether pay-per-episode or free, all of these new delivery approaches have profound consequences for broadcasters, cable companies and satellite-providers as their decades-long models of program distribution increasingly shift to the Internet.

The Internet and Broadband

An early vision of the Internet was provided by Vannevar Bush in a 1945 article for *Atlantic Monthly* magazine.[26] Bush, an engineer and administrator known for his role in the development of the atomic bomb, envisioned a device called *memex* that could electronically store books, records and other information for quick access. Memex could be accessed via a desk with translucent screens, a keyboard and a set of buttons and levers. With several screens, one set of material could be left in position while another was called up. Furthermore, the material would be cross-referenced via electronic links and branches.

Bush's device was never built. But two decades later, psychologist Joseph C. R. Licklider played a seminal role in what would become the Internet. In 1962, Licklider was appointed head of the Information Processing Techniques Office (IPTO), a division of a federal agency called ARPA (Advanced Research Projects Agency). Stung by the Soviet Union's pioneering launch of satellite *Sputnik* in 1957, the US government formed ARPA to develop technology with military applications. To advance this mission, Licklider enthusiastically promoted the use of computers for collaboration among his researchers in what he called an "intergalactic network". The network could be used for sharing files and programs, but it might also exchange "electronic mail." In 1968, Licklider and Robert Taylor wrote an article called "The Computer as a Communication Device" and boldly stated that "in a few years, men will be able to communicate more effectively through machine than face to face."[27] Government researchers and administrators such as Ivan Sutherland, Robert Taylor and Lawrence Roberts subsequently operationalized Licklider's vision and developed what would become ARPANET. The first ARPANET message was exchanged between computers in Los Angeles and Palo Alto, California, on October 29, 1969. ARPANET grew rapidly and a total of 14 nodes were in operation by 1971.

ARPANET was the first computer network to use *packet switching* – a system of breaking up messages into segments (packets) and sending them over the network in short bursts via multiple routes. The concept of packet switching was revolutionary. Prior to ARPANET, all communication networks used *circuit switching* in the same manner as telephone calls. Once the communication link was established the circuit would stay open the whole time, utilizing all the available bandwidth even if no information was being sent at a given moment. Packet switching more efficiently utilized network resources by dynamically allocating bandwidth in such a way that short-burst computer traffic would take up bandwidth only when data were actually being transmitted. It permitted multiple messages to share the same line and users could communicate with many computers at the same time.

The next major advancement was the 1974 invention of a new set of communication protocols (the rules by which computers talk to one another) called *transmission control protocol/internet protocol or TCP/IP*. Invented by Vinton Cerf and Bob Kahn, TCP/IP allowed computers to communicate with each other regardless of what hardware and software they might use. By 1982, all the pieces were in place for what became known as the *Internet.* Though the term "Internet "was coined by Cerf and Kahn in 1974, it didn't become widely used until the mid–1980s.

The Domain Name System (DNS) was adopted in 1984 to give computers on ARPANET a unique name (with suffixes such as .mil and .edu), instead of just a numerical IP address. A year later, the National Science Foundation started a network known as NSFnet to connect five supercomputing facilities, primarily to support university research. By 1990, responsibility for the Internet was transferred to NSFnet and ARPANET was dismantled. The National Science Foundation lifted the ban on Internet commercial activity in 1991 and the years that followed witnessed explosive growth. In 1995, private companies were providing Internet backbone services and NSFNET was officially decommissioned.

The so-called "killer application" of the Internet has to be the World Wide Web. Inspired by the contributions of Vannevar Bush and Ted Nelson (who coined the term hypertext), physicist Tim Berners-Lee invented the World Wide Web in 1989 while working at CERN, the European Particle Physics Lab in Geneva, Switzerland. Berners-Lee developed the Web's protocol (HTTP, hypertext transfer protocol) and the language of Web pages (HTML, hypertext markup language) as well as most of the pioneering communication software. Early Web clients were capable only of displaying text with very limited support for graphics. The first user-friendly graphical Web browser called *Mosaic* was created by Marc Andreesen, Eric Bina and a team of graduate students at the University of Illinois in 1993. *Mosaic* permitted images and text to appear together on the same page, while enabling the user to click on links with a mouse.

At the time, slow connections via telephone modems limited the types of media that could be effectively delivered via the Web. Text-based email downloaded quickly, but Web pages containing audio and video pushed the limits of human patience. Perhaps nothing has broken down the barriers to convergence as much as the development of *broadband* connections to the Internet. As Internet connection speeds have increased, *rich-media* products such as animation, audio and video have proliferated. The Federal Communications Commission defines broadband as any Internet connection with a data rate of at least 200 kilobits-per-second (kbps).[28] Today, broadband is most commonly piped into the home via telephone lines or a cable television system's coaxial cable, but it can be delivered by other methods including power lines and wireless connections. Broadband use is growing rapidly. A Nielsen/NetRatings study found that by 2006, more than 60 percent of online Americans connected to the Internet via a high-speed broadband connection –mainly through phone or cable lines.[29]

Thanks to broadband, telephone companies can deliver video and cable companies can bring us phone service. Both industries are invading the territory of the other, combining television, telephone and the Internet – bundling what they call a *triple play* of video, voice and data.[30] Phone company broadband goes by the name *Digital Subscriber Line* (DSL), while broadband delivered over cable lines is called a *Cable Modem*.

DSL

Alexander Graham Bell never envisioned using telephone lines to carry megabits of data or video. After all, the goal of the original telephone service was simple – to convey the human voice. The traditional phone service Bell created is still in use today and goes by the name *POTS*, or "plain old telephone service." It uses twisted-pair copper wires for transmitting voice telephone messages in analog form.

In the late 1980s, telephone companies began work on a technology called *Digital Subscriber Line* (DSL) aimed at carrying high bandwidth digital signals over their existing POTS infrastructure. The development and rollout of DSL in the late 1990s gave phone companies the ability to offer new services, while still providing analog voice messages. Theoretically, DSL can transmit data very quickly (up to 10 megabits per second), though the rate varies greatly depending on the length of the line and the gauge of the wire.[31] The rate actually delivered to most users is usually far less. Connections speeds are generally greater downstream than upstream (*asymmetrical bandwidth*) allowing users to receive data faster than they can send it. The bandwidth can be capped by the phone company and offered in packages that vary the amount of bandwidth available. In a typical DSL package, downstream speeds will be about 1.5 megabits per second, with upstream speeds in the range of 128 to 256 kilobits per second. However, *symmetrical bandwidth* service, where the upstream speed and downstream speed are identical, is also available. On the other hand, those living more than about 18,000 feet from a phone company's switching facility may not be able to get DSL service at all. The farther one lives from the switching facility, the slower the connection.

Telephone companies are also rolling out a new generation of television distribution called *Internet Protocol Television* (IPTV). IPTV is essentially a video-on-demand service that delivers television signals. For the telephone companies, this is accomplished through a broadband DSL connection. In one of the first IPTV arrangements, sister companies AOL and Warner Brothers began offering vintage TV shows such as *Welcome Back Cotter*, *Wonder Woman*, and *Kung Fu* free, with limited advertising, via their In2TV service.[32]

Work is underway to develop an improved protocol called *very-high-bit-rate digital subscriber line* (VDSL) that could transmit up to 50 megabits per second of data over existing phone lines. Such a technology could allow phone lines to support even high-definition television (HDTV).[33]

Cable modems

The other major player in consumer broadband access is the local cable company through *cable modems*. A cable modem is really not a modem at all, at least not in the traditional sense, because unlike old-fashioned telephone modems, cable modems are always on.

Cable systems deliver television by splitting the bandwidth on their systems into 6 megahertz channels. When devoted to Internet service, one of these 6 megahertz channels is used for moving data downstream, while another 6 megahertz channel carries upstream signals. In general, cable modem speeds are faster than DSL speeds, with a theoretical peak performance of 27 megabits per second. (compared with DSL's theoretical 10 megabits-per-second). However, the real-world speed of cable modems is in the range of 1.5 megabits-per-second. This is due primarily to the *shared bandwidth* aspect of cable modems, meaning that speeds decrease when many cable modem customers are online simultaneously. (DSL, with its *dedicated bandwidth*, isn't affected by a high number of simultaneous users). Like DSL, the bandwidth is usually asymmetrical, meaning the speeds are faster downstream than upstream. Unlike DSL, the data rate of a cable modem doesn't decrease the farther a home is from the company's head-end.

As telephone companies move to deliver video, cable is conversely invading the voice telephony market with a service called *Voice over Internet Protocol* (VoIP). Voice over IP technology allows the user to make telephone calls using a broadband connection instead of a regular analog POTS connection. DSL can be used for VoIP as well, but cable companies have taken the lead, seeing VoIP as a way to better compete with phone companies. VoIP also enables cable companies to distinguish themselves from competing satellite providers who are at a disadvantage in offering VoIP. The high *latency* of satellite services results in too much delay for effective real-time voice conversations. (Latency is the time it takes for an Internet packet to get from one point to another).

Instead of sending voice in a continuous, always open connection (*circuit switching*), VoIP works by digitizing the voice and sending it as Internet data in the form of packets – the system ARPANET introduced decades ago. The packets are essentially chunks of information sent in bursts across the Internet that are then reassembled on the receiving end and converted to sound. Consumers can utilize a regular phone, as long as it is connected to an adaptor which, in turn, is lashed to a broadband connection such as a cable or DSL modem. No computer is required, but users can talk from computer-to-phone or computer-to-computer if they desire. However, because modems are powered by household electricity, there's always the danger of power outages and disruption of service. In contrast, POTS phone service has its own power and continues to work during household electrical failure.

Some cable companies are providing unlimited local and long-distance VoIP phone calls, along with digital cable and broadband Internet service, for one flat fee. The services work with existing phones and users can keep their current phone numbers. The packages also offer call waiting, caller ID, call forwarding and voice mail features – just like traditional telephone companies. At the start of 2006, the cable industry's VoIP customer count was already at 2.4 million homes.[34]

Cable also has been offering video-on-demand (VOD) for a number of years. As of late 2005, the number of cable customers who watched on-demand TV doubled from the year before – to 23 percent – and this activity is expected to continue to mushroom.[35] Like the telephone industry and its DSL service, cable companies are working to increase the amount of cable modem bandwidth. The cable industry envisions devoting additional channel capacity to more efficiently deliver VOD and other services and hopes to provide a blazing 160 megabits-per-second in the near future.

Broadband over Power Line (BPL)

Even utility companies are entering the broadband marketplace by using the power lines that come into our homes. The service, called *Broadband over Power Line* (BPL), superimposes radio waves onto electrical signals and transmits data to the home at speeds equivalent to cable modems and DSL via existing electrical wires. Some BPL systems deliver the signals all the way into a home. Other BPL systems utilize WiFi (see below) connections to carry Internet data from the pole outside the house to the PC inside. Once inside the home, the Internet connection is available at any wall outlet – giving the term "plug and play" new meaning. Power lines may be the most pervasive utility infrastructure in the world, so BPL could be rolled out where there is no existing telephone or cable infrastructure. There are about two million miles of distribution power lines in the United States alone.[36]

BPL is capable of high *throughput* (the amount of data sent during a given period of time) and low latency, making it good for real-time applications such as gaming. However, adapting power lines for high-speed data transmission is not easy. Problems with power transformers, surge arrestors and network interference have hindered BPL development. Its use of radio frequencies to carry data also makes BPL a potentially large source of unwanted interference for short wave and ham radio signals. Unlike phone wires or cable's coaxial cables, power lines are unshielded and radio waves have a tendency to migrate outside the wire. Broadband over Power Line is still largely experimental, but roll-outs are underway in a number of cities and outlying areas.

Wireless broadband

WiFi (Wireless Fidelity) connections offer Internet users high bandwidth on local area networks without wires, so are quickly becoming an attractive alternative to wired connections. WiFi is the generic term for a standard certified by the IEEE (Institute of Electrical and Electronics Engineers). It comes in different versions such as IEEE 802.11b, which is capable of up to 11 megabits-per-second, and IEEE 802.11g, which can deliver speeds up to 54 megabits-per-second. Both standards use radio frequencies in the range of 2.4 gigahertz for communicating, placing them in the same part of the spectrum as microwave ovens and cordless phones – products that can consequently interfere with WiFi networks and compete for bandwidth. WiFi access points or "hotspots" have a physical distance limitation of between 100 feet and 300 feet. They can be set up in businesses, homes, airports, hotels, coffee shops or anywhere desired. In fact, some communities have installed enough access points to cover an entire neighborhood[37] or town.[38]

A next generation WiFi standard known as 802.11n will provide even faster connections with transfer rates up to 540 megabits-per-second – ten times faster than 802.11g. The standard would also increase the operating distance from a hotspot. An even newer standard called *WiMax* (Worldwide Interoperability of Microwave Access), or IEEE 802.16, will provide speedier connections to much larger geographic areas. A single WiMax base station could serve a large cluster of businesses, or hundreds of homes, thus simplifying broadband wireless deployment. A mobile version of WiMax is also under development.

EV-DO (Evolution Data Optimized) is a broadband wireless data service, offered by mobile phone companies, and capable of delivering video feeds to a cell phone. EV-DO gives cell phone users a high-speed Internet connection anywhere they can get a signal, even in cars and trains. The application is useful for more than just cell phones, however. With an EVDO card, a computer can access the Internet anywhere within a cellular phone company's coverage area. No more searching for a WiFi hotspot. If you can pick up a cell phone signal, you're on the Internet with a high-speed connection that is approximately 10 times faster than traditional cell phone connection speeds. Companies such as Sprint and Verizon have been rolling out EV-DO very rapidly with the goal of having their entire networks capable of EV-DO in the near future.

Another form of high-speed wireless connectivity is provided by *satellite-based broadband*. Satellite Internet access is a good option for rural areas without cable television or for locales beyond the reach of DSL. The first generation of satellite Internet systems were one-way only, providing data from satellite to home, but requiring a slower phone line and modem for sending data. Today, two-way satellite

Internet services provide downstream speeds in the range of 1.5 megabits-per-second. While the overall throughput is similar to that of other broadband technologies, satellite Internet suffers from poor latency (in this case, the delay inherent in getting data from one point to another via satellite). This is due to the 44,480-mile path to and from the satellite that the signal must travel, producing a latency of a second or more. For applications such as online gaming that require minimal delay, satellite delivery may be a poor choice. In the future, low-orbit satellites or even solar-powered airships could reduce this latency problem.

Whether using wires or wireless connections, it is clear that enough collective bandwidth exists to provide voice, audio, video and interactive services via a variety of avenues.

Internet Applications

The Internet has transformed business and social behavior, while undergoing almost constant innovation and change. In the 1990s, Internet services expanded from basic email to the World Wide Web, and video streaming. Today's growth is in applications that enable sharing and relationship building, while allowing users to both customize and contribute to the communication experience. Such applications as blogs, wikis, RSS news feeds, podcasts and vodcasts are challenging broadcasters and other traditional businesses to better engage their audiences, users and clients. In this section, we'll look at a few of the Internet applications driving this change.

Peer-to-peer file sharing

P2P has become a common and controversial Internet activity. At the beginning of the twenty-first century, Napster mounted the first widely-used system for sharing music files over the Internet. It became extremely popular, especially among college students with fast Internet connections. A big part of Napster's appeal was the fact that the sharing software was free and music could be downloaded without charge, even though much of it was copyrighted. Such massive copyright abuse led the music industry to take action and Napster was eventually shut down as a free service.

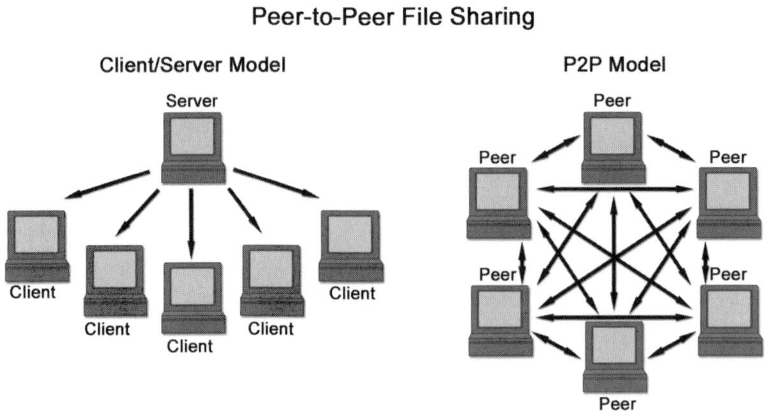

Peer-to-Peer File Sharing

Figure 6.2 Two file-sharing systems. Left: Some file-sharing systems utilize a "client/server model" for tasks such as maintaining file lists and managing connections. Right: A true "peer-to-peer" file-sharing system is more efficient because each computer has the same capability. Users exchange files directly with no central server managing the network.

Although it was referred to as a peer-to-peer file sharing system, Napster didn't technically fulfill the requirement of peer-to-peer because it relied on central servers to maintain song lists and manage the connections. Centralized file sharing systems are inherently inefficient and it wasn't uncommon to experience slow connections and long download times when many files were being exchanged. The centralized approach also made it possible to terminate the service by simply shutting down the master servers.

In true peer-to-peer networks, users share files directly with each other and there is no centralized server or database (see Figure 6.2). The first of this generation of P2P clients, Gnutella, was released in early 2000. An improvement to the approach appeared in 2001 with the release of the FastTrack P2P protocol. FastTrack is more commonly recognized by the names Kazaa, Grokster and Morpheus – all clients that connect to the FastTrack Network. FastTrack allows faster searches and download speeds than earlier P2P networks, especially when many users are connected. Like other P2P networks, the creators of client programs like Kazaa have been taken to court by the music publishing industry over infringement of copyright protected works.

FastTrack-based clients such as Kazaa have been popular, but the introduction of *BitTorrent* provided the first practical way to download very large files, such as television shows and movies. BitTorrent's growth has been astonishing with one study estimating it accounts for up to 35% of all Internet traffic.[39] BitTorrent works especially well for downloading files that are in high demand and therefore exist on many Internet-connected computers. The BitTorrent client breaks each user's file into smaller parts, typically less than a megabyte in size. When attempting to acquire a file, a computer analyzes the connection speed of everyone who has the file and starts to download the pieces from those with the best connections. One piece may be obtained from User A, other pieces may come from User B, User C, and so on.

Once you gather pieces of the file, those pieces can be shared with others, even though you may not yet have the complete file yourself. Eventually, you'll find enough peers with all the pieces you need and can aggregate the complete video file. In essence, the more popular the file, the more likely the pieces will be available from fast connected computers. The system is excellent at distributing server load and bandwidth. But the fastest download speeds are given to those who share (upload) the most, a property known as *leech resistance*. This discourages "leeches" from only downloading files and not sharing with others.

BitTorrent has often been used for illegal exchange of copyrighted material. Where earlier P2P networks, such as Napster, were seen primarily as threats to record companies, artists and music stores, BitTorrent is most worrisome to those involved in television program and motion picture production and distribution. There are legitimate uses for BitTorrent. In late 2004, when people were looking for amateur video of the tsunami disaster in Asia, the demand was so high it became difficult for traditional media sources to handle the load on their websites. Many Internet users turned to BitTorrent to acquire high-quality videos that weren't easy to find or download elsewhere. Other legitimate uses include access to public domain video footage, independent films and music, open-source software downloads and game demos.

Blogs

A blog, or Web log, is usually nothing more than a web-based journal of a person's thoughts. The blog can be a running commentary on diverse topics, or focus on a specific subject such as politics, technology or music.

Blog entries are listed in reverse chronological order, meaning that the most recent entry shows up at the top of the screen. Blogs often contain links to other websites or other blogs and in some cases allow readers to comment on the original postings. This conversational aspect of a blog can lead to a *blogstorm* or *blog swarm* in which an eruption of opinion results in a large number of user postings. The process of creating and updating a blog can be automated via weblog software (*blogware*) to accelerate and simplify the posting process.

Television program producers, networks and local stations are incorporating blogs to enhance communication with their audiences. For example, Corey Miller, a writer for *CSI: Miami* uses a blog to provide insight on the show and maintain a relationship with loyal viewers of the program.[40] A number of NBC and MSNBC on-air personalities and journalists maintain blogs. Early adopters at the local station level included KCAL-TV (Los Angeles) which makes blogs available from its news, weather and sports personalities. Meanwhile, WKRN (Nashville) has assembled an inventory of local bloggers and makes these linkable via the station website. Such efforts are aimed at connecting to viewers who have turned away from traditional media outlets.

Wikis

A *wiki* is a linked set of Web pages that allows content to be written and edited collectively. Individuals can create or modify a Web page without review by an outside source. Most wikis are open for contributions from anyone, but may require contributors to first register for a user account. Wiki websites tend to generate a large amount of content and can be tremendously popular. However, the quality and consistency of the material often suffer due to the number of authors and their varying levels of subject knowledge. Inaccuracy and outright vandalism are constant problems. Most wikis will include style guides in an attempt to make the content more consistent from subject to subject.

Wiki sites are often encyclopedia projects. *Wikipedia* (wikipedia.org) is said to be the largest user-contributed encyclopedia by article and word count. The Wikipedia edit screen allows users to contribute their own text, edit text or even place hyperlinks via a simplified markup language that anyone can use. The site has spun off into several related projects such as Wiktionary, Wikitravel, Wikibook and Wikinews. It is licensed under the GNU Free Documentation License, which means that the content can be freely shared, copied and distributed, with or without modifying it, either commercially or non-commercially.

Wikis can exist on the local level as well. Cities such as San Francisco, California,[41] Davis, California,[42] and Calgary, Alberta,[43] have vibrant user-contributed wikis underway. Local media outlets may choose to start up and sponsor a community wiki project and encourage users to contribute. The collaborative nature of a wiki fits well with a broadcaster's mission to serve the public interest – fostering community interaction and providing an avenue for stations to reach viewers otherwise lost to the Internet.

RSS news feeds

An *RSS news feed* is an Internet technology for sharing selected news content from a website. In technical terms, it is described as a family of XML file formats intended for Web syndication. RSS can stand for either "really simple syndication" or "rich site summary." Websites that want to publish content such as news headlines and stories, can create an RSS feed for distribution. Employing special news reader software, RSS users subscribe to news feeds to receive a list of stories from a website or websites. The user can then click on a story headline or excerpt to be taken directly to the web page and the full story. For persons who visit a large number of news-oriented websites daily, the RSS news reader saves time because it will display updated content only on subjects those individuals have indicated are of interest to them. There's no need to visit multiple websites every day to check for updates because the software alerts the user when updated content becomes available.

Many traditional media outlets, including the major brodcast networks, cable networks and local broadcast stations, provide RSS news feeds. Often, a news organization will have multiple feeds based on reader interest subject areas. CBS News, for example offers feeds on Top Stories, US, World, Politics, Sci-Tech, Healthwatch, Entertainment, Business and Opinion.

Individuals or media outlets who want to publish content can create an RSS feed and arrange for distribution through what are called content *aggregators*. Aggregators allow Web users to build their own portals where they can customize content based on their interests. The system is most helpful to those who want to have updated content from multiple websites and display it all on one page.

RSS feeds are also useful within the blogging community. Individuals who read many blogs can use the news reader software to monitor their favorite blogs and show only previously unread entries. Major online news sites (such as Wired), news-oriented community sites (such as Slashdot) and portal sites (such as My Yahoo!) were early users of RSS. The RSS format is extremely versatile. RSS feeds can be used on a wide range of media devices such as PDAs (personal digital assistants), cell phones and email ticklers.

Podcasting

Podcasts, a combination of the words iPod and broadcasting, are audio programs, usually in the MP3 format, available for download via an RSS enclosed feed. Podcasting software applications, such as iTunes, check to see if new files are available and automatically deliver the content to the user. This changes the nature of downloading from a pull model (where the user has to go find out if new content is available) to a push model (where the software automatically pushes the new content to the user).[44]

Even though podcasts can be listened to on any personal audio player, the term itself illustrates just how ubiquitous Apple's iPod has become. While designed for users to transfer files to an iPod, podcasts can also be downloaded to a personal computer and listened to in applications such as Windows Media Player or iTunes. Podcasts can also include video material – what is referred to as *vodcasting*.

Using podcast software, a user subscribes to an RSS feed that lists available audio and/or video material from a content provider. There are number of software packages available including iPodder, Doppler, Primetime Podcast Receiver and iTunes. As with RSS news feeds, the software can recognize new content and provide a link for downloading. It will even automatically play the media in the chosen media player software. Podcasting is truly a many-to-many technology. Anyone can create self-published radio shows or music and make the content available to a potentially large number of people. According to Forrester Research, 12.3 million households will synchronize podcasts to their personal audio devices by 2010.[45]

Some media outlets are pursuing podcasting in order to expand their audiences. National Public Radio has become a leading podcaster and one of the most popular sources on the "iTunes Top 100." Much of their success is attributed to the creation of original content, instead of simply repurposing their existing programming. NPR has found that shorter content, in the range of 4–8 minutes, tends to be most popular, perhaps due to the multitasking nature of podcast listeners.[46] CNN Radio podcasts a number of feeds including the latest general and business news, along with ongoing CNN programs and special presentations. Most other major media outlets now podcast at least a portion of their existing content or provide material specifically aimed at podcast users.

In at least one case, podcasting is said to have "killed the radio star."[47] In May 2005, traditional over-the-air radio station KYCY (San Francisco) decided to convert to an all-podcast format. The AM station dumped its lineup of star syndicated talk show personalities and replaced it with podcasts from amateurs who submit their creations via the Web. Station executive Joel Hollander asserted that the new format "harnesses that power by serving our listeners with content developed by them for them."[48] Listeners receive the simulcast program on either 1550 AM or via the company's website and can submit their podcasts for inclusion on the station directly at that web address. The station's producers screen the submitted content to make sure it doesn't violate FCC rules and to assure that it meets quality standards.

Critics charge that, because of poor ratings, KYCY has surrendered the airwaves of a major market AM station to amateurs. But according to Hollander, "You have to make bets on new forms of technology – some work, some don't. We're making a bet that this might become the way people want to communicate."[49] Time will tell if the all-podcast format successfully replaces radio stars with radio amateurs – or just becomes more roadkill on the Information Super Highway.

Future Communication Technologies

The decades ahead will unveil technologies we cannot imagine today. Still, several experiments hint at what may be in our future. First, the development of even higher resolution video displays will result in *super high-definition systems.* What we call high-definition video today may be tomorrow's equivalent of NTSC. Japan Broadcast Corporation (NHK) has already demonstrated a *Super Hi-Vision* television system with a resolution of 4,320 lines (7,680 × 4,320 pixels).[50] The system uses 8 megapixel video digital video cameras and transfers data with no compression via a fiber-optic network. At 4,320 lines of resolution, Super Hi-Vision has more than 4 times the picture resolution of the sharpest high-definition picture in use today. While such technologies are ultimately aimed at the lucrative home/consumer market, their initial high costs will likely see them used first for medical, government and scientific applications that would demonstrably benefit from such pristine and lifelike imagery.

Many believe television will finally approach reality when it escapes its current two-dimensional constraints and ventures into the realm of 3-D. The prospect of *3-dimensional television* has long held our fascination. Popular culture has provided a conception of how it might look. The original *Star Wars* movie portrayed a flickering 3-D display of Princess Leia. With its holodeck, the television series *Star Trek: the Next Generation* got us to think about an immersive 3-D environment. We've even been able to experience it, thanks to those flimsy red-and-blue glasses, via such productions as *Captain EO* at the "Magic Eye Theater" inside Disneyland.

Volumetric displays render video images in a 3-D space, instead of on a traditional flat 2-dimensional surface. So called "3-Deep" displays are already emerging and may be commercially viable in the near future. There are two main technological approaches to creating 3-D displays. *Swept volume* uses either a projector or an array of lasers to bounce images off a fast rotating screen. The other approach, a *solid-state* system, employs a projector behind a stack of 20 liquid-crystal displays to create a solid image from a rapidly projected series of images.[51] Both of these systems create images that require no special eyewear and are viewable across a wide angle of view by several people in a room. But as with super high-definition displays, don't expect to see 3-D television in your home anytime soon. Look for it first in medical and scientific fields where their initial high cost will be justified.

The satellite industry is pursuing its own technological marvels. While the idea of placing lasers in space sounds like the story line from a spy movie, laser or *optical satellites* could one day produce very high-speed voice, data and video connections. Already under development by the military and defense researchers, *laser communication* or *free-space optics* exploits a portion of spectrum near the range of visible light (infrared), but still invisible to the eye. Lasers can be digitally encoded to carry voice, data and video at extremely high speeds. The communication is received via telescopic lenses that collect digitally encoded photon streams. The data rate of optical satellites might one day achieve a level of performance comparable to ground-based fiber technology in the range of 10 gigabit-per-second. Such systems would allow high-speed communication with lightweight satellites providing broadband Internet to even the most remote regions of the planet.

Chapter Rewind

Satellite radio is competing with terrestrial AM and FM stations by offering hundreds of usually commercial-free channels. Its providers offer nationwide coverage from satellites and ground-based repeaters with programming available in homes, cars and on small portable receivers. For traditional radio broadcasters, HD (high-definition) radio is a new technology that allows transmission of multiple channels of near CD quality audio, along with song information and on-demand interactive services all in the same amount of spectrum previously devoted to conventional broadcasting.

A confusing array of production, transmission and display standards poses a challenge to the adoption of digital television. Broadcasters can choose to produce and transmit programs in a variety of formats. Signals are often up-converted or down-converted from a program's native resolution to be transmitted and shown at other resolutions. For consumers, purchasing a digital television set requires knowledge of a set's display and tuner capabilities. Depending on how the signal will be acquired, a set may also need additional hardware. Digital compression technology allows cable and satellite companies to transmit more channels, but the practice can lead to material degradation – loss of quality associated with compression.

Using spot beam technology, DBS satellites can now provide customers with the signals of their local television stations. With dedicated receivers and dishes, even high definition versions of these station broadcasts are available in some markets.

Device convergence involves the integration of personal communication, computing and consumer electronics on a single appliance or delivery system. Broadcast networks, local stations, cable companies, satellite services, land-based and mobile phone entities, as well as Internet companies are all seeking to deliver multiple programs and interactive features. In many cases, former competitors are striking deals to cooperate and expand program distribution options. Consumer frustration over incompatible technologies led to the formation of the Digital Living Network Alliance (DLNA). Comprised of computer, consumer electronics and mobile communication companies, the DLNA is working to develop standards and create interoperable devices.

Broadband Internet, delivered through cable modems and the telephone industry's DSL, is enabling a "triple play" of voice, data and video. Phone and cable providers are invading each other's territory with phone companies offering Internet protocol television (IPTV) and cable companies providing voice over IP (VoIP) phone service. Wireless broadband services are connecting neighborhoods and towns, while EV-DO allows mobile phone companies to deliver broadband to phones and computers wherever a cell phone signal can be acquired.

The first packet-switched computer network was known as ARPANET and it eventually grew into what we know today as the Internet. Two events led to tremendous growth in Internet use – the 1989 invention of the World Wide Web and the 1991 decision to allow commercial activity on the Internet. Broadband connections have spawned a variety of controversial peer-to-peer (P2P) file sharing applications on the Internet. P2P has grown to include decentralized systems capable of efficiently exchanging very large files such as popular television shows and movies. Internet applications such as wikis, blogs, RSS news feeds and podcasts have allowed users to both customize and contribute to the communication experience, challenging traditional media to engage consumers before they are lost to the Internet.

Future communication technologies may include super high-definition displays, 3-dimensional displays and optical satellites capable of blanketing the planet with incredibly high-speed communications services.

SELF-INTERROGATION

1   How do the methods of satellite program delivery used by XM satellite radio and Sirius satellite radio differ from each other?

2   How could HD radio allow AM stations to better compete with FM stations?

3   Describe the dispute between broadcasters and cable companies with regard to material degradation of digital television programming.

4   How are direct broadcast satellite companies able to carry local broadcasting signals to their home markets without having to send out thousands of stations to the whole country?

5   What are the various functions now offered on hybrid devices, such as mobile phones?

6   What are the three categories or "islands" of digital media products and what is being done to advance interoperability among them?

7   What does it mean to use broadband for a "triple play" and what kinds of companies are competing to offer triple play services?

8   What is VoIP and how is it different from POTS telephone service?

9   What are the advantages and disadvantages to using Broadband over Power Line when compared to DSL or cable modem service?

10   How is EV-DO different from Wifi or WiMax and what types of devices utilize the technology?

11   How does packet switching differ from circuit switching?

12   Why is BitTorrent an effective system for sharing large files that are in high demand?

13   What are blogs, wikis, RSS feeds and podcasts and how are traditional media outlets utilizing them?

14   What are laser satellites and how do they differ from traditional satellites?

NOTES

1   Consumer Electronics Association, "Household Penetration of CE Products Soars in 2005"; 17 May 2005; available from http://www.ce.org/Press/CurrentNews/press_release_detail.asp?id=10753; accessed 26 December 2005.

2   Matthew N.O. Sadiku, "Satellite Communications," in Mohammad Ilyas (ed.) *The Handbook of Ad Hoc Wireless Networks* (Boca Raton, FL: CRC Press LLC, 2003), 8–21.

3   Yuki Noguchi, Washingtonpost.com, "XM Radio to Offer Voice Commands as New Service"; 30 December 2005; available from

http://www.washingtonpost.com/wp-dyn/content/article/2005/12/29/
AR2005122901434.html; accessed 30 December 2005.

4   Ted Schadler, Forrester Research, "The Future of Digital Audio";
available from http://www.forrester.com/Research/Document/Excerpt/
0,7211,36428,00.html; accessed 24 September 2005.

5   Michael Kanellos, C|Net, "High Definition Radio Gears Up for
Reality"; available from http://news.com.com/High-definition+radio
+gears+up+for+reality/2100-1041_3-5722285.html; accessed 20
October 2005.

6   Susan Whitall, The Detroit News, "Brave New Waves: High-Definition FM
Radio Ups the Ante for Traditional and Satellite Formats"; available from
http://www.detnews.com/2005/screens/0509/30/E01-332480.htm;
accessed 27 November 2005.

7   Apple Computer, Inc., "Apple iTunes for Your Mobile Phone"; available
from http://www.apple.com/pr/library/2005/sep/07rokr.html; accessed
14 November 2005.

8   3G, "World's First Walkman Mobile Phone"; 2 March 2005; available from
http://www.3g.co.uk/PR/March2005/1124.htm; accessed 28 December
2005.

9   EVDOinfo.com, "What is EVDO?"; available from http://www.
evdoinfo.com/EVDO/Info/What_is_EVDO?_2005021237; accessed 27
November 2005.

10   MSNBC, "X Marks the Spot for Giddy Gamers"; available from
http://www.msnbc.msn.com/id/10099496; accessed 22 November 2005.

11   Andrew Brandt, PCWorld, "First Look: Sony's Impressive PlayStation
Portable"; available from http://www.pcworld.com/reviews/article/0,aid,
120117,00.asp; accessed 22 November 2005.

12   Wired News, "Playing Games With Sony's Player"; 7 April 2005; available from http://www.wired.com/news/games/0,2101,67151,00.html;
accessed 22 November 2005.

13   Antone Gonsalves, TechWeb, "Microsoft, Cable Industry Agree on Cable-
Ready PCs"; 16 November 2005; available from http://www.techweb.
com/showArticle.jhtml?articleID=174300552; accessed 22 November
2005.

14   MSNBC, "Intel to Partner with TiVo on PC Platform"; 30 November
2005; available from http://www.msnbc.msn.com/id/10264362/from/
RS.3; accessed 30 November 2005.

15   Digital Living Network Alliance, "Overview and Vision White Paper";
June 2004; available from http://www.dlna.org/about/DLNA_Overview.
pdf; accessed 8 January 2006.

16   ZDNet, "NBC, CBS to Offer Shows for 99 Cents"; 7 November 2005;
available from http://news.zdnet.com/2100-1040_22-5938773.html;
accessed 22 November 2005.

17 James A. Martin, Washingtonpost.com, "ITunes Video to Go: Download Video to Your Notebook to Watch On-Screen or on Your TV"; 18 November 2005; available from http://www.washingtonpost.com/wp-dyn/content/article/2005/11/17/AR2005111700154.html; accessed 22 November 2005.

18 Saul Hansell, The New York Times, "NBC to Sell TV Shows for Viewing on Apple Software"; 7 December 2005; available from http://www.nytimes.com/2005/12/07/technology/07apple.html; accessed 7 December 2005.

19 Linda Moss, Multichannel News, "DirecTV's Pre-Emptive Strike"; available from http://www.multichannel.com/article/CA6297835.html; accessed 9 January 2006.

20 Yuki Noguchi, Washingtonpost.com, "CBS to Make a Soap For the Smaller Screen"; 12 January 2006; available from http://www.washingtonpost.com/wp-dyn/content/article/2006/01/11/AR2006011102283.html; accessed 12 January 2006.

21 Yuki Noguchi and Greg Schneider, Washingtonpost.com, "Courting the Web Giants"; 7 January 2006; available from http://www.washingtonpost.com/wp-dyn/content/article/2006/01/06/AR2006010601892.html; accessed 10 January 2006.

22 TiVo, "ToVoToGo Transfers"; available from http://www.tivo.com/4.9.19.asp; accessed 22 November 2005.

23 May Wong, Associated Press, "Yahoo-TiVo to Blend TV, Web Services" 7 November 2005; available from http://www.siliconvalley.com/mld/siliconvalley/13102550.htm; accessed 22 November 2005.

24 Rik Fairlie, C|Net, "Slingbox Puts TV Content on Your Notebook"; 28 December 2005, available from http://www.cnet.com/4520-10602_1-5619284-1.html; accessed 28 December 2005.

25 Marguerite Reardon, C|Net, "NBC to Air Nightly News Online"; 1 November 2005; available from http://news.com.com/NBC+to+air+Nightly+News+online/2100-1025_3-5926934.html; accessed 22 November 2005.

26 Vannevar Bush, "As We May Think," *Atlantic Monthly* (July 1945), 101-8.

27 J.C.R. Licklider and Robert W. Taylor, "The Computer as a Communication Device," *Science and Technology* (April 1968), 21.

28 Federal Communications Commission, "Federal Communications Commission Releases Data on High-Speed Internet Access Services"; 22 December 2004; available from http://hraunfoss.fcc.gov/edocs_public/attachmatch/DOC-255669A1.pdf; accessed 15 November 2005.

29 Nielsen/NetRatings, "Broadband Internet Use Grows in the U.S."; 28 September 2005; available from http://www.netratings.com/pr/pr_050928.pdf; accessed 15 November 2005.

30   Converge Network Digest, "Blueprint: Telco Triple Play"; available from http://www.convergedigest.com/bp-ttp/index.asp; accessed 15 November 2005.

31   Ann E. Barron, Gary W. Orwig, Karen S. Ivers and Nick Lilavois, *Technologies for Education: A Practical Guide* (Greenwood Village, CO: Libraries Unlimited, 2002), 166.

32   The News & Observer, "AOL, Warner Bros. Welcome Back Shows You Can View on Your Computer"; 15 November 2005; available from http://www.newsobserver.com/104/story/367263.html; accessed 15 November 2005.

33   Charles H. Ferguson, *The Broadband Problem: Anatomy of a Market Failure and a Policy Dilemma* (Washington, DC: The Brookings Institution, 2004), 34.

34   Alan Breznick, Cable Digital News "North American MSOs Count 2.4 Million VoIP Subscribers"; 1 December 2005; available from http://www.cabledatacomnews.com/dec05/dec05-1.html; accessed 29 December 2005.

35   Ken Belson, The New York Times, "Fiddling with Formats While DVD's Burn"; 25 December 2005; available from http://www.nytimes.com/2005/12/25/technology/25cnd-format.html; accessed 26 December 2005.

36   Leonard A. Sagan, *Electric and Magnetic Fields: Invisible Risks?* (Palo Alto: CA, Electric Power Research Institute, 1996), 14.

37   Winston-Salem Cityscape, "WiFi on 4th"; available from http://www.cityofws.org/wifion4th; accessed 28 December 2005.

38   Bob Brewin, Computerworld, "Michigan City Turns on Citywide Wi-Fi"; 30 July 2004; available from http://www.computerworld.com/mobiletopics/mobile/wifi/story/0,10801,94928,00.html; accessed 28 December 2005.

39   Adam Pasick, Yahoo! News, "Livewire – File-Sharing Network Thrives Beneath the Radar"; available from http://in.tech.yahoo.com/041103/137/2ho4i.html; accessed 30 December 2005.

40   CBS.com, "Writer's Blog"; available from http://www.cbs.com/primetime/csi_miami/blog.php; accessed 31 December 2005.

41   San Francisco Wiki; available from http://sanfrancisco.wikicities.com/wiki/Main_Page; accessed 31 December 2005.

42   Davis Wiki; available from http://www.daviswiki.org; accessed 31 December 2005.

43   Calgary Wiki; available from http://calgary.wikicities.com/wiki/Main_Page; accessed 31 December 2005.

44   Michael Silbergleid, Television Broadcast, "Why Aren't You Podcasting? Because You're Smarter Than That"; available from http://television-broadcast.com/articles/article_1054.shtml; accessed 20 October 2005.

45 Schadler, "The Future of Digital Audio"; accessed 24 September 2005.

46 Mark Glaser, USC Annenberg Online Journalism Review, "Will NPR's Podcasts Birth a New Business Model for Public Radio?"; 29 November 2005; available from http://www.ojr.org/ojr/stories/051129glaser; accessed 31 December 2005.

47 Xeni Jardin, Wired News, "Podcasting Killed the Radio Star"; 27 April 2005; available from http://www.wired.com/news/digiwood/0,1412,67344,00.html?tw=wn_tophead_1; accessed 24 September 2005.

48 Benny Evangelista, SFGate.com, "KYCY-AM First Station to Convert to All-Podcast Format"; 28 April 2005; available from http://sfgate.com/cgi-bin/article.cgi?f=/c/a/2005/04/28/BUGETCGGLH1.DTL; accessed 24 September 2005.

49 Jardin, "Podcasting Killed the Radio Star"; accessed 24 September 2005.

50 Yoshiko Hara, EETimes, "Japan Demonstrates Next-Gen TV Broadcast," 3 November 2005; available from http://www.eetimes.com/news/latest/showArticle.jhtml?articleID=173402762; accessed 4 November 2005.

51 Alan Sullivan, "3-Deep: New Displays Render Images You Can Almost Reach Out and Touch," *IEEE Spectrum* (April 2005).

CHAPTER 7

Content Challenges

Peter B. Orlik

Given the variety of media technologies now available, there are more opportunities for the development of new content types and packages than ever before. Such an environment presents electronic professionals with increased competition as well as an ever widening spectrum of programming decisions that must be made with ever greater precision. This chapter explores some of the issues and quandaries these decision-makers face.

The Consumer-Centric Cosmos

The overarching reality shaping today's electronic media universe is that technology has empowered the consumer as never before. "In the digital era," observes former Hewlett-Packard CEO Carly Fiorina, "the future is one in which consumers watch or listen to what they want to watch, when they want, at any time they want, on any device. This is a generation that will not wait for content to be delivered to them at a prescribed time."[1] So not only must content be appealing, it must be appealingly served up in a way that meshes with the target consumer's lifestyle and patterns of daily existence.

We are progressing toward what advertising executive Bob Greenberg labels "a new world of ubiquitous content on demand. One of the biggest shifts in marketing and advertising today is the move from the outbound world of media-driven campaigns to the inbound world of information on demand."[2] As we learned in Chapter 3, the twentieth-century model for electronic media content featured mass communicators pushing content to audiences waiting to receive it. In 1929, over half the US radio audience tuned in at 7 p.m. every week-night to hear *Amos 'n' Andy*. Twenty years later, a similar percentage of the television audience reserved their Tuesday evenings for Milton Berle's *Texaco Star Theater*. These early successes set a broadcasting pattern that endured for the rest of the century. Different shows, but the same pattern, emerged as other countries rolled out their own broadcasting systems. The broadcaster selected the show, scheduled a time for it to air, and then utilized increasingly sophisticated promotion campaigns to entice audiences to tune in at the appointed time.

The advent of the VCR wedged the first small crack in this "outbound" monopoly by providing the opportunity for viewers to time-shift programs to more convenient times – assuming they could master the machine's instruction manual. DVR's like TiVo have simplified but not fundamentally changed this time-shifting process. The media companies push the content out and the consumer watches it in "real time" or records it for more convenient consumption later.

But newer digital media go much farther than this as the previous chapter indicated. Different yet mutually-supportive versions of the same program property now are *cross-platform*. People can access it at their leisure on television, the Web, their mobile phones, and even imbedded in the video games they play. At the same time, consumers can repackage such content for themselves through podcasts and blogs. In addition, the slightly more technologically hip can add to the mix content they themselves produce to construct their own audio and video webcasts.

It might be assumed that such intense personalization will totally supplant the traditional corporate media structure. However, such self-created efforts lack a feasible business model. They do not possess the resources to either structure or effectively market even niche-appeal programming. Therefore, the threat to mainstream media each of these one-person activities pose is minor. It is not that they individually will steal audience, but that collectively, they may divert a measurable percentage of the total listener- or viewership from traditional media consumption. Of much greater concern to broadcasters and other professional communicators is that they themselves will not possess the cross-platform capability to deliver their programming in ways that mesh with consumer preferences.

As mentioned earlier, the "inbound" world is governed by people's individualized demands for content – not on when and how communications corporations decree that content will be delivered. "The end-user is getting greater control by the day over the type of information and style of information they're getting," states Ketchum Advertising/Chicago's managing director Paul Rand. "If the mainstream media is one of the tools they use, then that's terrific. But increasingly, we're finding the value of these broader-based outlets is diminishing because of the increased ability for people to personalize where and how and in which format they're getting their information."[3] Cable executives like Comcast CEO Brian Roberts are sensitive to that same trend. "We look at TV changing to personalization," Roberts reveals. "When you look at [Web super search engine] Google, people get what they want more than ever before. We have to be there first with that experience."[4]

This does not mean that the traditional program schedule mode that built local stations and broadcast and cable networks is going to vanish in the foreseeable future. But it is no longer the only way through which consumers can choose to be entertained and informed. As we have discovered, there are now three screens through which an audience can be reached and aggregated in a size sufficient for a viable business: the TV, the computer, and the cell phone. And these screens can complement – or compete with – radio too. Therefore, established media enterprises like Disney's sports network ESPN are staking out space on every available delivery platform to make it as easy as possible for their customers to find and spend time with them.

The challenge is to create audio or video content that takes advantage of viable technologies without penalizing consumers who prefer one platform over others. That requires taking a much more comprehensive view of what programming today is all about. NBC/Universal Television Network Group president Randy Falco has said that his company can no longer consider itself a broadcaster, or cablecaster but instead, a programmer of all sorts of media: "I really don't believe that in the 21st century, any media organization can be organized around silos. People are finding new ways to consume media."[5]

So despite the technologies employed, the business of electronic media is still very much about content; very much about the programming that attracts consumers and thereby the advertisers seeking to reach them. Companies strive to own as much of this content as they can and to leverage it in as many ways as possible. This is not an entirely new trend. Recall that CBS chairman William S. Paley began building his television network in the late 1940s by luring top personalities from competing radio networks and then giving both them and his own established radio stars cutting edge exposure via the emerging TV.

Contemporary Audio Programming

Having just alluded to radio history, it is appropriate here to survey audio's current content characteristics. Notice we use the term "audio" rather than "radio" in order to encompass sound communication enterprises beyond conventional broadcasting.

Firmly following the pattern begun with the 1950s localization of radio, most aspects of today's audio communication business remain format intensive. Whether over the broadcast air, on the Web, satellite-beamed, or called up on a cell phone, professionally-generated audio entertainment and information are packaged to deliver a carefully circumscribed content to a well-defined target audience presumed to find that content attractive. A well-conceived and promoted format provides a particular group of listeners with what they want and expect – and thereby serves up this definitive type of audience to advertisers or other financial supporters most interested in reaching such people.

As a whole, the radio industry has tended to embrace formats designed collectively to appeal to the 25–54-year-old age group – because the bulk of advertising dollars are targeted to reach consumers somewhere within that range. The most popular US and Canadian radio formats (see Table 7.1) clearly reflect this targeting. Such demographic clustering clearly under serves both youth and senior citizens. However, it does open up audio opportunities to attract them on the Web. Stations can create streams separate from the one that carries their on-air signal in order to offer edgy alternative music services for youth, or "beautiful," big band or classical fare popular with seniors. Webcasters who do not own a radio station can do precisely the same thing. The geographic reach and comparative low-cost of Web station operation make these programming streams economically attractive to

Table 7.1   The dozen most popular radio formats in the US and Canada based on number of stations airing each

Format	Number of stations
Country	2288
Christian/Religious	2063
Adult Contemporary	2024
News/Talk	1329
Oldies	1213
Sports	952
News	787
Talk	771
Classic Rock	718
Contemporary Hit/Top 40	717
Spanish	654
Rock/AOR	627

advertisers beyond those preoccupied with 25–54 delivery and provide the Web programmer with a much greater potential for profitability than broadcast airing of those formats. Satellite radio, with its capacity of over 100 channels, and a business model that depends more on listener subscription fees than advertising revenues, can serve all of these audiences simultaneously. However, satellite service lacks the local character that a radio station can exploit.

In fact, *localism* is radio's strongest competitive advantage. Despite all its "emerging media" competitors, the radio station can still compete for those choice-empowered consumers with programming attuned to a particular community. "Look at your stations," radio consultant Bill Suffa advises managers. "Are you really local? Do you know who your listeners are? . . . The competitors are further behind on the content side and are just walking toward the starting gate on localism. They can do it, and they very well may try, but radio can give them a reason not to do it."[6]

Projecting a sense of localism depends most on community-oriented air personalities. Unlike voicetrackers who are recording on-air banter for multiple locations, truly local talent convey information and references that are of particular relevance to the geography around them. Granted, voicetracking is an economical way to bring top air talent to multiple stations. However, it fails to take advantage of radio's most potent weapon against competing audio media. "Good beats 'local' every time," concedes Audience Development Group's president Tim Moore. But, he adds, "Good-local, always beats good-national."[7]

Consequently, developing on-air talent is as big a challenge for radio stations as it is for the newer audio services seeking to lure away listeners. Adept air people also provide a programming advantage that distinguishes professional communications businesses from personal recording devices. "Personalities are a major part of creating radio's overall entertainment package," asserts Dave Lange of consulting firm McVay Media. "I can't imagine how you'd recreate that with an iPod."[8] Satellite radio programmers also recognize that compelling personalities are at least as important to their success as they are to the success of conventional stations. But because the satellite operators are charging for what radio stations provide free, their talent must be even more attractive.

This is not just a matter of luring top radio commodities like Howard Stern to satellite – because there are just too many channels to fill. Instead, pioneer "*satcasters*" like XM and Sirius soon learned that they needed to breed a new brand of personality as well. Lee Abrams, who helped build the FM industry through his introduction of AOR (album oriented rock) and successor formats, was hired by XM to work similar magic for that subscription satellite service. Personalities have become a central part of his focus. Abrams maintains:

> We are forced to rethink everything and deliver a sound that is different, fresh, and compelling enough to be worth every cent. We installed a 'cliché' buzzer. If our staff comes up with a 'been there, done that' cliché, they are buzzed. Three buzzes and you're fired! The point is to liberate ourselves from everything we've learned and start over again building a sound geared for the new millennium listener."[9]

This, of course, means knowing that listener; plugging into that consumer-centric cosmos discussed in this chapter's first section. "We have spent 90 percent of our time focusing on the sending end, instead of the receiving end," says consultant Tim Moore. "Few talent can profile their listeners beyond age demos."[10] Today's professional programmers need to help talent with this profiling task in order to solidify bonds with their listenership. Certainly, many consumers will turn to iPods and similar devices for instantaneous access to their favorite audio features. But well-packaged and delivered audio *programming services* – whether on radio, satellite, or the Web – can refresh what these favorites might be. "The iPod is all about music you already own," industry analyst Ross Rubin maintains, "and radio is often about music exploration and exposes people to music they might like. Ultimately, I think consumers want both."[11]

A Video Programming Overview

As is the case with radio, the old "television" and its programming are being stretched and reshaped by multiple new delivery technologies. Broadcast TV stations must now coexist with cable, DBS, and both wired and wireless video streaming. In some instances, such as cable and satellite, the station's main channel is simply picked up and distributed in its entirety. In other environments, like the Web and wireless cell phone system, key extracts from or supplements to the station's over-the-air signal are utilized to help promote that signal. At the same time, the conversion to digital television means that the station itself can generate several program streams simultaneously and become a "multicaster." But while the station can benefit from all of these extensions of its reach, so can competing stations. And cable, DBS, Web, and other wireless enterprises are not just retransmitting broadcast stations, but are distributing content of their own. All of this makes for an extremely competitive environment in which content must be effectively packaged, promoted, and differentiated.

For years, there has been talk of the "500 channel universe" and how this would completely fragment what had been the "television" business. What research has discovered, however, is that most consumers tend to visit no more than 20 channels – and devote most of their viewing to only 4 or 5 of these. Of course, all viewers are not regularly visiting the same 20 services. So the trick is to aggregate a blend of occasional viewers with regular ones in sufficient numbers to justify costs being charged to advertisers, subscribers, or redistributors. This aggregation must be accomplished through programming that satisfies viewers' desires in a cost-effective manner despite the multiplicity of alternatives these viewers can access. In the 1960s, when there were only three networks available, network executive Paul Klein advanced his *LOP* (least objectionable program) *Theory* which proposed that the show disliked by the fewest number of viewers would win. Unquestionably, this theory has been turned on its head in today's multi-channel, consumer-centric environment.

But while the three-choice-limited world of the 1960s is long gone, the bulk of US television programming is still controlled by a relative handful of sources. The majority of the seven media conglomerates discussed in Chapter 5 have gathered program production, broadcast and cable network ownership, and local station operation into vertically integrated enterprises. Although the shows they produce are sometimes sold to other companies, each tends to air its own studio's creations because this is more profitable than licensing programming from competitors. Currently, about 70 percent of broadcast network-aired shows are produced within that network's parent company. A result of this integration is that there is pressure for a network to stay with a show produced by its sister studio longer in order to accumulate enough episodes (usually 100) for the property to be licensed for rerunning on individual stations via lucrative off-net syndication. "When it's yours, you tend to be more patient," observes media buyer Stacey Lynn Koerner.[12] Subsequent to network run, the company's own group of *O & Os* (owned and operated stations) are prone to pick up the show, thereby giving an initial boost to its syndication rollout across the country.

A program can also be bounced back and forth between the conglomerate's broadcast and cable networks, making more money off a hit series, stretching revenue on a marginal one, or occasionally, "growing" a property on cable before upgrading it to a broadcast network run. A variant on this strategy is for a sister cable network to rerun the previous year of a series for which the broadcast network is still licensing new episodes, The hope is that this will refresh interest on the part of previous broadcast viewers of the series while attracting new ones via cable exposure.

All of this means that the products of the big media conglomerates have a much greater opportunity to reach consumers than do the products of unaffiliated producers. This applies to programs made for *first-run syndication* (not previously network-aired) as well as reruns of network fare. Despite King World Production's great success in making and marketing such hits as *Wheel of Fortune, Jeopardy!* and *The Oprah*

Winfrey Show, even this company found going it alone less viable and is now part of CBS Corp. This gives King World properties an advantage, if needed, in placement on the CBS O & O line-up. Still, there are several station group owners such as Tribune and Gannett outside the "Big Seven" conglomerates who also control multiple major market outlets and therefore have clout in negotiating for popular syndicated programs. Even a "Big Seven"-owned studio will think twice in selling to a sister O & O group if it can get a better deal from an outside station or list of stations.

Stations in larger markets each spend $100,000 or more per week to license their inventory of syndicated shows. In addition, they often must turn back half the commercial time within these shows to the syndicator who sells this time to national advertisers through a process known as *barter*. "That is fine as long as the product delivers, but not many do," points out Dick Haynes, senior vice president with media consulting firm Frank N. Magid & Associates. "Stations can't continue to spend money on something that doesn't work."[13] Consequently, in a move mirroring what radio stations are doing, local television stations are returning to their roots and producing more local programming. Not only does this save them money on expensive syndicated fare but it also differentiates them from other stations and video delivery systems.

Many of these local productions, like *Colorado & Company* on Denver's KUSA and *Show Me St. Louis* on that city's KSDK are lifestyle offerings airing in daytime hours. But some other outlets use local shows to counter competitors' syndicated fare in late afternoon or *prime time access* (the hour before the start of network evening programming). Together with locally-produced news, these in-house productions are a way to court viewers and local advertisers (who sometimes sponsor entire show segments) in the most community-relevant manner possible.

Video News Issues

News has always been a mainstay of local television programming, of course, and the tactical need for localism has made it even more important. Stations which had reduced or entirely abandoned news production have recently reinstated it – especially after purchase by a larger group that sees the competitive necessity for this type of programming. Technological advances and the reduction of trade union jurisdictions mean that fewer news personnel can do more jobs, reducing the high station payroll costs that had always been associated with news production. Reporters now shoot and edit their own pieces as well as voicing them. They can even *repurpose* the story for Web or radio distribution as a means of cross-promotion and advertising revenue enhancement.

Well-focused local news can be a significant profit center for a station and a major enhancement of its market identity, but it has also been criticized for its over-emphasis on the flashy, the sensational and the trivial. When combined with the "if it bleeds, it leads," mentality, an explosion of computer graphics and a preponderance of human interest "feel good" pieces make for a mishmash of infotainment rather than incisive information. Thus, the challenge for the station news director is to keep the fast pace and shorter story characteristics designed to attract younger viewers while still conveying stories of substance and import to the community. It is difficult to withstand the pressure to trivialize the news when the ratings of competitors doing so are rising, especially among advertiser-sought demographics.

Nor is broadcast *network* news insulated from such pressures. As their audiences skew older and they find themselves up against the 24-hour news cycles of cable news channels and the Web, the broadcast nets are rethinking the way they do news and searching for glossier presentation styles that are no longer built around "voice of God" anchors.

Changes must be made. For between 1994 and 2004, the dollars generated by the broadcast news units of ABC, CBS and NBC dropped from 10.5 percent of their nets' total revenue to 4.8 percent and the combined shares of their evening news programs declined from a 52 share of audience to a 37 share.[14] Part of this continuing decline is due to the greater spectrum of choice available to viewers. But part is also the fault of a format that seems stodgy and irrelevant to younger consumers. If the broadcast nets are not to abandon news altogether, they must find a way to build a more attractive package with diminished budgets while hopefully still upholding a tradition of reputable journalism. Viewers will "always stop to watch the car wreck, low behavior by people in high places, volcanoes erupting or hurricanes coming ashore," said NBC anchor Tom Brokaw shortly before his retirement, "but they'll also stop to watch the complicated story about governance, foreign policy, science and the economy if you explain that it's important, why it's important, and present it in a way that engages the viewer."[15]

For its part, cable news can benefit from its 24-hour status – particularly when big stories are breaking. On the other hand, cable news viewership drops substantially during slow news times making for programming that can be both thin and repetitive. One solution is to replace some of the conventional newscasts with interview and "talk shows" similar to what the radio news/talk format has done for decades. For further differentiation from competitors, a news channel can modify the scrupulous balance historically expected from television news and instead, feature commentators and commentary that espouse a more partisan ideology – as Fox News has done.

While this approach is certainly controversial, the controversy itself can build viewership. And its defenders submit that partisan journalism has long been a feature of the newspaper industry in both the US and other countries. Because cable is not licensed to local communities as are local stations and because people choose to subscribe (as with newspapers), cable news programming should be allowed wider latitude say its supporters. The key factor is whether a channel is straightforward about the thrust of its coverage. "Fox isn't exactly pursuing a stealth strategy; anyone who can't figure out that it's in the tank with the Republican Party must be brain dead," observes *New York Times* critic Frank Rich. "It's more insidious when some of its more fair-and-balanced competitors blow-dry the news not to serve an ideology but to tell the public what they think the public wants to hear."[16]

In addition, some would argue that the news and pseudo-news found on the Web has forever blown out the narrow attitudinal boundaries within which electronic journalism historically functioned. Webcasts created by organizations unaffiliated with any conventional news medium abound and blogs of every political stripe carry "news" that can be the product of a single, sometimes unidentified, individual. Though such activities individually lack the reach and penetration of broadcast and cable, mass media journalists increasingly find themselves having to react to what is being reported over the Web as they search for confirmation or contradiction of what suddenly high-profile Webcasters are saying.

The Incredible Sports Hulk

One part of the electronic news spectrum that has always been presumed to be highly partisan is the looming sports sector. Home town media are expected to favor (or at least emphasize) home town teams and even some national commentators are permitted to express their preference for one team over another. Because of its powerful and deep-seated appeal, sports programming is also hugely important to electronic media programmers. It can generate significant ratings and thereby serves as a potent platform from which to promote the station or network as a whole.

Paradoxically, however, most local newscasts are reducing the time allotted to sports, burying sports coverage at the end of the show where it receives no more than 4 minutes of airtime and sometimes as little as 120 seconds. (Stations try to put the best face possible on the latter by having their sportscaster call it "The Two Minute Drill".) The reason for such abbreviated coverage is that avid sports fans are no longer going to wait for the end of a late evening news show to get the scores. They have already gotten them from cable or the Web where they may have watched or listened to an entire game. The sports segment is typically even shorter on the early evening newscasts which air before most games are started and skew to a female demographic that, though more interested in sports than the past, prefers other content at the dinner hour.

This is not to say that stations are abandoning sports. If they are able to secure broadcast rights for a local team's games, this can be a huge image-builder for the outlet – provided such coverage does not preempt too many network shows. This is why many of the stations still airing games are independents. They do not have to worry about accommodating network schedule patterns and the games fill time for which these "indys" would otherwise have to purchase expensive syndicated shows. But more broadcast networks mean less independent stations and thus, fewer of these over-the-air outlets to convey local games. Nevertheless, even network affiliates can keep loyal fans tuning them in by emphasizing coverage of local teams – even minor league and high school ones – while downplaying mention of sports outside their market. If the station traditionally has a strong sports image and a respected sports reporter, it can additionally treat sports as mainstream news by moving that reporter's stories into the main body of the newscast alongside those filed by business, city hall and other "franchise" journalists.

Most station executives recognize that actual game telecasts and other end-to-end events coverage are now principally the province of regional and national networks – especially cable ones. Scheduling issues and escalating rights costs make such a pattern inevitable. For broadcast networks, coverage of high profile major league and major college games, though expensive, can still bring in significant revenues from male-seeking advertisers and provides tie-in opportunities to promote the network image and other programming across several platforms. It can even be used to drive viewership to appropriate fare on the parent company's cable nets. Many games do not bring in sufficient advertising dollars to offset their high rights and production costs. However, network executives believe that the residual value to them and their affiliates in carrying the games makes for overall profitability. And given the prominence of sports in the culture, a network without any significant sports presence is taken less seriously. It was FOX's 1994 outbidding of CBS for National Football League rights, for example, that brought that network to parity with the previous "Big 3" in the minds of potential affiliates, advertisers, and viewers.

Nevertheless, even broadcast network sports has taken a back seat to cable and DBS. Regional channels like those operated under the Fox Sports umbrella and national channels like ESPN can devote 24 hours a day to sports and market themselves accordingly. This enables them to cluster large amounts of major and minor sports programming and pass the costs on to advertisers, cable or DBS operators and, in pay-per-view situations, consumers themselves. Without the need to conform to affiliate station schedules or to provide a wide spectrum of general interest programming, cable and DBS sports purveyors can leverage their sports exclusivity to create healthy profits and pass on the cost of rights increases to others. They can further repurpose their well-recognized programming on the Web and over the "third screen" (as ESPN is beginning to do via its own branded cell phone service).

The costs of sports programming may be high, but the primacy of sports in attracting hard-to-reach young males in particular solidifies it as a "must buy" for several key advertiser categories. The real-time nature of sporting events also makes coverage of them far less susceptible than most program fare to time-shifting and other consumer delayed recording activities that dilute the value and timeliness of the adjoining advertisements. Finally, improvements in audience measurement techniques may heighten the value of sports programming even more. Traditional methodologies do not count out-of-home TV viewing such as that which occurs in bars, health clubs and college dorm lounges – the very places where people are likely to be watching sports in group settings. But portable people meters (PPMs), worn like a pager or cell phone, are now being rolled out that read messages encoded into the audio signals of all types of media. When carried into these sports-heavy viewing environments, PPMs are likely to give a significant boost to sports ratings and consequently, to the revenues generated by sports programming.

The powerful appeal of sports has been exploited in virtually every country as a means of building interest in emerging electronic media systems. Many innovations in program production and transmission have been linked to the coverage of sporting events. The continuing challenge is how to cope with the expense of this coverage while not restricting broad-based consumer access to it.

Reality Vehicles and Product Placement

Like sports coverage, *reality* programming is the depiction of seemingly unscripted actuality for the purpose of viewer entertainment. Unlike expensive sports, "reality" frequently is embraced because of its comparative *low* cost of production. As compared to scripted entertainment, reality programs are not saddled with the financial burden of high-priced stars, highly paid writers, and high-end studio sets. Reality fare can be shot "on the fly" using lower-end cameras, to capture sketched out (though scrupulously formatted) situations performed by on-camera amateurs. It can be produced quickly, in any number of episodes, and because of reality's comparatively low cost, employed equally well during the regular viewing season or as a summertime fill-in series. "You don't have a tremendous amount of flexibility with a predominantly scripted schedule of 22 episodes per show," WB network CEO Jordan Levin points out. "Reality is a tool that allows networks to schedule year-round, with fresh, relatively inexpensive programming."[17]

However, reality shows are also properties that may be widely popular for a brief moment and then disappear with little residual economic value. Unlike successful scripted series, reality shows seldom last long enough to accrue the 100 or so episodes that make them viable vehicles in syndication. Even if they did, most reality programs do not repeat well because the nature of *American Idol, The Apprentice, Survivor* and most others is a series-spanning contest that culminates in the anointing of a winner or team of winners. Like a televised football or hockey game, there is not much interest in re-watching once the final score is already known.

So despite the sudden popularity that a reality series can sometimes achieve, it is largely disposable programming that, some would claim, does not contribute to the long-term vitality of the channel that carries it or the television medium in general. On the other hand, it can also be argued that money the networks save by scheduling reality fare can then be redirected to support the development of scripted projects that might otherwise not be fundable. Whatever the case, young viewers in particular have embraced reality even as they find scripted television less and less attractive. This preference is certainly not lost on network executives or the top producers of reality programming who are licensing customized domestic versions of their shows in countries around the world. In fact, the first reality breakthroughs on US television were concepts born in Europe.

To hold these younger viewers and broaden show impact, reality producers are also concentrating attention on the interactive component of their properties. Companion websites can offer game spin-offs of the program's central contest, facilitate chat rooms for fans to share observations and commentary and even provide online applications for people seeking to participate in next season's episodes.

In virtually every developed country, people born since 1980 have never known a world of limited media options. They therefore use such interactivity to control (more overtly than their parents ever did) what they perceive as a constantly mutating media experience. "The emphasis today must be on immediacy, a recognition that popularity is fleeting and celebrity disposable," observes Betsy Frank, executive vice president for research and planning for MTV. "This is just what the top reality shows deliver to a group of people who associate 'forever' with 'yesterday'. These shows are created to burn big, burn bright and then disappear."[18] The most successful leave behind profits for their producers and the channels that carry them – plus a formula that can be regenerated into a subsequent series.

Reality series have also led the way as vehicles for *product placement* – the integration of a branded commodity within the program's very fabric. While the practice goes way back to the stilted *integrated commercials* of 1930s radio and 1950s television, today's reality shows do this in a much more sophisticated way by sculpting the entire entertainment concept around the client product or products. When producer Mark Burnett had *Survivor* challenge winners receive bags of Doritos and six packs of Mountain Dew as rewards, mere product placement became *branded entertainment* to launch a new era in "zap-proof" advertising.

Reality shows have become the perfect branded entertainment environment because "audiences expect a degree of commercialism in reality shows," argues Campbell Mithun media buying executive John Rash. "There isn't a suspension of disbelief as within fictionalized dramatic or comedic characters. And so audiences are willing to accept Coca-Cola on *American Idol* and Chrysler Crossfire Roadster on *The Apprentice*, because in many ways those programs are in themselves commercialized for the pop industry and Donald Trump and his enterprises."[19] The reality genre has caught on at just the right time, adds Patrick Quinn, president of research firm PQ Media, "because the success of shows like *Survivor* and *The Apprentice* have convinced more marketers that product placement is often a strong supplement to the deteriorating effectiveness of the 30-second spot."[20]

This rush to find alternatives to easily avoided commercials means that product placement is not confined to reality shows. CBS Corp. chief executive Les Moonves has predicted that by the end of the decade, as much as 75 percent of all prime-time broadcast *scripted* series will indulge in the practice.[21] Meanwhile, on the daytime soaps, a perfume has been created out of an *All My Children* storyline, General Motors' Onstar global positioning system was scripted into episodes of *The Young and the Restless* and *General Hospital*, and Butterball turkeys roasted on *As the World Turns* and *Guiding Light*.[22] Through the practice of *virtual placement*, products can even be digitally inserted into previously-produced series, with unbranded boxes and cans now acquiring a specific brand's label and other products showing up on what had been empty countertop, desk, and coffee table space. All that is needed for the insertion is an original shot at least six seconds in length.

Such ventures may well prove a financial safety net for electronic media that see their production costs rising and traditional advertising revenues threatened by new commercial-skipping technologies. But there are also dangers associated with product placement and its more embedded branded entertainment cousin. When content is developed with an advertiser's needs in mind first, audience interests inevitably take a back seat. Viewers have a certain tolerance for placements that are integral to the storyline and add a real-world texture to depicted environments. But when the placement is done in a blatant, jarring, or disjointed manner, it can undermine the consumer's relationship with the brand, the show, and even the outlet that carries it. That, in a consumer-centric world, can be fatal.

The Ethnic and Global Dimension

Reality programming and product placement are not just US phenomena but are practices easily shared via a global media economy. As previously mentioned, some of the most successful US reality fare had its genesis in European programs. Concepts that prove popular in one country are now imported to others – particularly in an era when international co-production is becoming more common as a means of sharing costs and reducing risks. This means that US electronic media cannot afford to underserve an increasingly diverse audience at home as well as abroad.

Ethnic audiences in the United States are growing rapidly in both numbers and economic clout. The Census Bureau predicts that, in 2050, non-Hispanic whites will become the minority. Programmers and marketers are not waiting until then to respond; particularly as buying power among heretofore "minority" groups is outpacing that of the historic racial majority (see Table 7.2).

This growth is especially pronounced among Hispanics and the media that have arisen to serve them. Newcasts and selected other programming on the two major Hispanic broadcast networks, Univision and NBC Universal-owned Telemundo, regularly win key demographic ratings races in large markets like New York, Chicago, Miami and Los Angeles as well as in smaller markets with large Hispanic populations. As Table 7.1 demonstrated, Hispanic radio is similarly prominent. In

Table 7.2 US ethnic group buying power. Move diversified spending creates the climate for greater diversity in programming. Figures in billions of dollars.

	1990	*2000*	*2002*	*2007*	*% Growth vs. 1990*
White non-Hispanic	$3,739	$5,800	$6,252	$7,910	112
Hispanic	233	491	581	926	297
African-American	317	559	646	853	169
Asian	118	255	296	455	286
American Indian	19	36	41	57	200
Other	85	198	234	379	346
Total	4,511	7,339	8,050	10,580	135

Source: Reprinted with permission from the July 7 issue of *Advertising Age*. Copyright, Crain Communications Inc. 2003.

addition, as early as 2004, no less than 75 cable television channels targeted Spanish-speaking consumers in the United States.

But the electronic media's ethnic expansion is not limited to the Hispanic sector. African-American-oriented radio formats and cable television channels have long been a part of the programming scene even though advertisers were slow to recognize and reward this presence. Asian services are now starting to come into their own as well with a relatively small number of radio outlets but an increasing prominence on digital cable and satellite. In 2005, for example, DirecTV announced it would distribute three of MTV World's channels in the United States: MTV Desi, for the South Asian community, MTV Chi for Chinese and MTV K for viewers of Korean descent.

Today's US electronic media much more accurately reflect the multiculturalism of their country and world than when a pair of White men portrayed two Blacks on *Amos 'n' Andy*. But the continuing challenge is to bring program and facility resources to parity with the needs of the communities to be served – and to do this in a way that is integrative rather than exclusionary. The pulsating global marketplace, stimulated by wide-footprint technologies like satellite and the Internet, makes it feasible to share entertaining and insightful programming across ethnic, national and continental boundaries. But it will take skilled and responsible professionals to capitalize on this opportunity in a way that makes both business and society-building sense.

Chapter Rewind

Technology has empowered the electronic media consumer as never before. People are no longer content to passively wait for the outbound transmissions from media outlets. They now choose the time, shape and composition of their programming. Radio has become part of a larger audio arena structured to appeal to carefully segmented audiences. Stations accomplish this through precise formatting while satellite radio utilizes 100 or more separate channels. Localism, executed by skilled on-air talent, is the most important tool a radio station possesses to differentiate itself from competing outlets and technologies.

Television is also being reshaped by emerging delivery technologies that allow the station to extend its signal and provide multiple services – but also expose it to a wide array of new competitors for viewers' time. The seven major vertically-integrated conglomerates control significant elements of US program production as well as broadcast and cable distribution, aggregating more content choice into fewer corporate hands. As with radio, localism is a way for a TV station to stand apart. News coverage is a significant component of this localism but is susceptible to charges of superficiality in its struggles to reach younger viewers. Broadcast network news must also find ways of attracting the younger demographic while cable's 24-hour news channels have to discover methods for meaningfully filling their time during slow news periods. Some are utilizing more news/talk features and taking a more partisan approach in order to differentiate their product from that of competitors. Meanwhile, blogs and other unconventional Web information sources continue to push back electronic journalism's traditional boundaries.

Sports programming is hugely popular, but its coverage within station newscasts is being reduced as fans are finding other ways to receive this content more quickly and comprehensively. Stations can continue to serve sports fans by concentrating on local teams. Broadcast networks must secure game rights, even though they are costly, in order to be taken seriously by major affiliates and advertisers. However, cable and DBS all-sports networks enjoy significant advantages in sports coverage because they can pass on rights costs to systems or subscribers and can leverage their vast amount of sports content across several delivery platforms.

Like sports, reality programming is very popular while avoiding sports' and scripted content's high production costs. Reality series do not generally repeat well but provide scheduling flexibility, significant program "buzz" (particularly among younger audiences), and interactive content opportunities. Reality properties are also prime vehicles for product placement and the more sophisticated branded entertainment because the show concept itself can be sculpted around client products. Scripted shows are exploiting placement techniques too as a means of providing sponsors with alternatives to commercials that viewers can skip. The global media economy transplants these and other techniques from country to country and encourages risk-sharing co-productions. At the same time, a multicultural program outlook is essential within countries as well. In the US as elsewhere, the growth of ethnic audiences' size and economic clout makes program diversification essential.

SELF-INTERROGATION

1. Distinguish between "outbound" and "inbound" media worlds.
2. What are the factors that have contributed to the onset of media consumer-centrism?
3. What is the radio industry's prime demographic target and why? What alternate vehicles exist to serve listeners outside this target?
4. What can radio and television stations do to fully realize their localism advantage?
5. Explain the LOP Theory and why it is no longer relevant.
6. To what new factors are station and network broadcast news operations having to respond?
7. Why is sports such an important program element and yet, a decreasing part of local television newscasts?
8. What is a PPM and how might it help bolster the economics of sports programming?
9. From programming and production standpoints, what are the advantages and disadvantages of reality series compared to scripted ones?

232   *Peter B. Orlik*

10   How does *branded entertainment* differ from simple *product placement*?
11   What is the significance of the year 2050? How are the elements comprising that significance impacting electronic media programming now?

NOTES

1   Steven Meyer, "Like It or Not, It's an iPod World," *Radio Ink* (April 11, 2005), 32.
2   Bob Greenberg, "FutureVision, Cont.," *Adweek* (May 23, 2005), 24.
3   Matthew Creamer, "Personal Tech Becomes Priority for PR,: *Advertising Age* (June 13, 2005), 4.
4   Daisy Whitney, "Hot Wired," *Advertising Age* (June 6, 2005), S-6.
5   John Higgins, "Universal Helps NBC Keep Its Balance," *Broadcasting & Cable* (May 2, 2005), 12.
6   Bill Suffa, "The Path to the Future Goes Through 'Competitive Media'," *Radio Ink* (February 21, 2005), 9.
7   Tim Moore, *Audience Development Group Confidential White Paper* (May 26, 2005), 5.
8   Abbey Klaassen, "iPod Threatens $20B Radio-Ad Biz," *Advertising Age* (January 24, 2005), 57.
9   Michael Keith, "Lee Abrams – From Broadcast to Satellite," *Journal of Radio Studies* (December 2004), 223.
10 Moore, *Audience Development*, 5.
11 Klaassen, "iPod Threatens $20 Radio-Ad Biz," 57.
12 J. Max Robbins, "Faulty Forecasts," *TV Guide* (December 20, 2003), 28.
13 Allison Romano, "The New Wave of Homegrown Fare," *Broadcasting & Cable* (January 24, 2005), 24.
14 Claire Atkinson, "TV News Can't Carry Own Weight," *Advertising Age* (January 17, 2005), 28.
15 Tom Brokaw, speech at the Radio and Television News Directors' Association Edward R. Murrow Awards (New York), October 4, 2004.
16 Frank Rich, "Happy Talk News Covers a War," *nytimes.com* (July 18, 2004).
17 Eric Schmuckler, "Facing Reality," *Adweek* (May 31, 2004), SR33.
18 Betsy Frank, "Check Out Why Young Viewers Like Reality Programming," *Broadcasting & Cable* (July 7, 2003), 7.
19 John Rash, "It Can Be Overdone," *Adweek* (May 17, 2004), SR42.
20 Marc Graser, "Product Placement Spending Poised to Hit $4.25 Billion in '05," *Advertising Age* (April 4, 2005), 16.
21 John Consoli, "The Word on Placement: It's Following the Script," *Adweek* (July 26, 2004), 8.
22 Paige Albiniak, "The Place to Place a Product," *Broadcasting & Cable* (February 14, 2005), 18.

CHAPTER 8

Regulatory Challenges

Louis A. Day

In the twenty-first century the bright line distinctions among media are rapidly disappearing. In fact, the observation that telecommunications technologies and media are converging has already become a cliché. In the future, such familiar terms as "television," "radio," and "telephone" will be irrelevant as the differences between the various delivery systems recede. In the process, however, we should not forget the historical narrative of the regulation of electronic media during the last century.

History's Legal Lessons

History has much to teach us if we will only listen. Unfortunately, the wisdom of the ages frequently goes unheeded as we struggle to fashion public policies to manage the introduction of new technologies and other scientific advances. Our fascination with technology has at times become pathological and often leads to unrealistic expectations and predictions. The introduction of each new technology is accompanied by a fresh array of futurists and pundits whose reach often exceeds their grasp. Or as Jeffrey Abramson and his co-authors remind us in *The Electronic Commonwealth*, history has been "notoriously unkind" to predictions about the media. The tyranny of prophecy can lead us into the "trap of technological determinism."[1] The fact is that technology is simply a means for delivering content, but it cannot control the political, economic, and social environment that must embrace its promise of change and cultural progress.

History teaches us that technological prophecies, particularly in the formative stages of any new medium, tend to coalesce around the extremes of utopianism and skepticism. Consider, for example, the development of cable television. Those who came of age in the 1960s can still recall the optimistic prognostications about cable, including, among other things, the end of the traditional workplace in favor of an "electronic cottage," a wired nation in which the two-way interactive capability of cable would permit viewers to talk back to their TV sets, and the availability of one hundred channels of programming by the early 1970s. Of course, the first two predictions yielded to the pragmatism of the cultural marketplace, and it took several decades for cable to approach 100-channel delivery.

On the other hand, television has previously defied many skeptics to emerge as the most influential mass medium of the twentieth century. Early television "picture tubes" like Vladimir Zworykin's iconoscope (see Figure 8.1) were seen as little more than industrial

Figure 8.1   Vladimir Zworykin and his prototype iconoscope. Source: Photo courtesy of David Sarnoff Research Center.

curiosities. Even a decade after the first practical demonstration of television, a doubting industry executive speculated that the problems of interference and the costs of producing and transmitting programs would deter any widespread adoption of the new medium. And despite impressive exhibition of television at the 1939 World's Fair, a *New York Times* reviewer still concluded that the average family would not have time to sit with their eyes glued on a TV screen.[2]

It is commonplace observation that American society is in the middle of a *communications revolution* as evidenced by the shift from an industrial to an information economy. Futurists predict a wide variety of "future shocks" to the cultural infrastructure,[3] but the term "revolution" denotes a radical break from the past, and the more pragmatic and cautious media analysts note that new media are really extensions of old media. Change is *evolutionary*, not *revolutionary*. The foundation of the Internet, for example, is computer technology, which had been around for decades prior to the cultivation of cyberspace.

If the "past is prologue," both old and new electronic media will pose regulatory challenges that are already familiar to us. Considering the well-entrenched commitment to the traditional electronic media (e.g., radio and television), thinking outside the regulatory box may be difficult. But new media realities require it. Before examining the legal implications for the new media, a brief survey of some of the most pressing technological issues is in order.

Beyond Broadcasting: Laws for New Media

The technologies of the twenty-first century are likely to produce an unprecedented communications explosion. These technologies will both improve upon old media and contribute to the development of new electronic media. Because of such a drastic change in the telecommunications environment, government regulators are already contemplating the shape of communications policy for the next several decades. The energizing force behind the new communications revolution is digital technology. The digital age is one in which information gets stored, processed, retrieved, and transmitted in a format that computers can handle. As we have discussed, *digitalization* has found its way into television, radio, satellite communications, telephone, and cable. Digital technology has also hastened the convergence of several electronic media venues: digital television, digital satellite radio, satellite television, broadband cable, and the Internet.

Digital television

As noted earlier, television licensees have until 2009 to convert from the traditional analog system to the much higher quality digital system. From a regulatory standpoint, this interval during which the conversion is supposed to take place has implications for cable. Some stations may transmit both digital and analog signals, but cable operators have argued they do not have the channel capacity to carry both. Thus, the FCC appears to be inclined to require cable and satellite systems to carry the "primary video stream" from each broadcaster. The determination of which signal is primary would be left to the broadcaster.[4]

Digital satellite radio

While the FCC has had to contend with technological advances in television, the radio industry has undergone a technological revolution of its own. Also known as *digital audio broadcasting* (DAB), digital radio is the audio counterpart to digital television and its most impressive manifestation, HDTV. DAB's primary advantage for listeners is the sound quality. In 2002 the FCC approved a system that allows for the simultaneous transmission of analog and digital signals. The ultimate goal, of course, is the digitalization of the entire radio industry.

Digital technology has also facilitated the development of *satellite radio*. The FCC first allocated spectrum space for satellite radio in 1992, but the new service did not make its official debut until 2001 (two weeks after the 9/11 terrorist attacks) when XM Satellite Radio went on the air with more than 100 channels offering a diverse menu of music, sports, news, and talk. Sirius Satellite Radio launched a competitive service the following year. From a regulatory standpoint, this development might have been of only passing interest. However, as early as 1997, in anticipation of the imminent debut of satellite radio, the FCC announced that satellite radio providers should abide by the political broadcasting rules (described in Chapter 4) while not being bound by other FCC content regulations such as those governing indecency. This differentiation from traditional radio licensees is based upon the fact that satellite radio does not use the public airwaves and is a subscription service akin to cable. It was for this reason that "shock jock" Howard Stern, who was responsible for millions of dollars of indecency fines against his broadcaster employers, abandoned over-the-air broadcasting for the more libertarian venue of satellite radio.

Satellite television

Satellite TV providers are currently engaged in an intense competition with cable for the hearts and minds of subscribers as evidenced by the unabashedly aggressive commercials emanating from both camps. However, satellite television began operations at a competitive disadvantage. Under the terms of the 1988 Satellite Home Viewer Act, satellite companies were prohibited from providing subscribers with signals from local television stations. TV customers, therefore, were inclined to stick with cable to have access to their favorite local stations.

The Satellite Home Viewer Improvement Act of 1999 corrected this competitive imbalance by allowing satellite companies to carry local stations to viewers living in the stations' dominant market area (DMA) as defined by Neilsen Media Research. However, if a satellite provider selects this option, it must carry all local stations in the market that request carriage, including non-commercial stations.

Like satellite radio, DBS providers must abide by the political broad-casting rules of the Communications Act. In addition, the FCC declared in 2000 that satellite TV must follow the network nonduplica-tion and syndicated exclusivity ("*Syndex*") rules described in Chapter 4. Satellite services also are subject to a "Sports Blackout Rule." If a local TV station is not carrying a local game, then "distant" stations carrying the game may not be shown in the local blackout zone.

Broadband cable

Since its genesis, cable television has evolved steadily from a local retransmission pipeline into a sophisticated and technologically complex avenue offering a wide variety of services and program content. Competitively, cable has progressed from a locally monopolis-tic enterprise to one that must compete with other multi-channel providers such as satellite television. Digital technology has made pos-sible not only higher quality transmissions and a greater diversity of cable programming, but it has also opened the door to a variety of services targeted toward well-defined audiences. Viewers are increasingly in the driver's seat when it comes to content selection. One commentator described this cable renaissance this way:

> Video-on-demand is like a digital video store, which allows subscribers to select from a wide range of programs and events for viewing at their own convenience. Near-video-on-demand allows customers to view programs with start times every 15 or 30 minutes. Additional interactive provisions include services such as electronic program guide, video games, home shopping, interactive commercials, teleconferencing and transaction processing, personal travel planning, and other multi-media on-line applications.[5]

But despite the development of cable as a multi-dimensional medium, including the capability of providing telephone and Internet service, for the foreseeable future most consumers will treat cable primarily as an entertainment and information medium. In this context, it poses the same regulatory challenges concerning content and "must carry" that the FCC has grappled with for several decades.

The Internet

On a global level, the dominant technology of the third millennium is undoubtedly the Internet. Whether one considers "the Net" to be a truly new medium or a platform for converging old and new media content, the transformation occasioned by the appropriation of the Internet for popular usage has been profound. The Internet is the first electronic medium to be truly democratic in the sense that everyone with access to the Net can produce content for audience consumption on an international scale. The nature of the net insulates it from most direct government regulation, but as we shall soon see, there have been some attempts to regulate the most objectionable forms of content, such as indecency. In addition, because of the ubiquitous nature of the Internet it poses unique legal challenges.

Convergence and the Legal Landscape

We live in an era of unprecedented technological change as new media rapidly become old media. Media practitioners of the future must be nimble in their capacity to adapt to this new environment. Video, satellite, digital, and computer technologies have become so commonplace that we no longer think of them as fads or novelties. They are the indispensable instruments of the information age. For this reason, there is a bright future for young, talented college graduates who wish to become a part of this growth industry.

The mantra for the media managers of the electronic media revolution is *convergence*. While media scholars and futurists do not entirely agree on a definition of this phenomenon, in essence convergence consists of the "coming together of computing, telecommunications, and media in a digital environment."[6] Convergence refers both to the merger of new media (such as Internet Service Providers like AOL) with traditional media enterprises *and* the blending of various types of media delivery systems, such as audio, video, and print. But for government regulators, convergence means that media-based regulations (i.e., treating media differently because of their unique technical qualities) may no longer be applicable. In the future, will it really make sense to hold broadcasters to a different regulatory standard than, cable, for example?

The convergence of new and old media has opened up possibilities that were unimaginable just a few years ago. For a number of years, as newspaper readership has declined because of competition from television and cable, publishers and editors have lamented the receding influence of their industry. But the Internet may have given them a new lease on life as newspapers have developed on-line editions. Originally, these technological experiments were little more than scaled-down versions of the print editions. But most papers of any size now provide real time updates of breaking news. Thus, they have become more competitive with their television counterparts. Similarly, the major news networks and many stations have their own Web editions that provide constant updates throughout the day and consumer-oriented information that expands upon stories featured on their regular news programs. In addition, some co-owned newspapers and television stations have "converged" their news operations so that reporters now produce stories for both print and broadcast. Such journalists, of course, must be cross-trained in print and electronic media, and some schools of journalism and mass communication have reconfigured their curricula to accommodate the industry's needs.

What does all of this convergence portend for electronic media regulatory policy? Legal issues of the near future are likely to cluster in three areas: (1) *ownership patterns*; (2) *media content*; and (3) *Constitutional (First Amendment) jurisprudence*.

For almost a century the media in the United States have moved persistently towards conglomeration and consolidation. The term "media empire" (or the more personified reference, "media mogul") was unheard of in the nineteenth century or even in the formative days of the broadcasting industry. But in recent years, chain ownerships and other media mergers, sometimes involving corporations with greatly diversified portfolios, have become commonplace.

The major broadcast networks, for example, are owned by such corporate giants as The Disney Company (ABC), CBS Corp. (CBS), and General Electric (NBC). Throughout the 1980s the Federal Communications Commission (FCC) encouraged media diversity through a policy of strict ownership limits. But the restrictions on the number of media properties a single entity can control have been steadily relaxed in the current deregulatory environment. For the most part, the 1996 Telecommunications Act allows individuals and companies to own as many radio stations as they wish and relaxes restrictions on the number of stations that a single broadcaster may operate in the same market. The consequences have been dramatic. In 1996, for example, Clear Channel owned just 36 radio stations. By 2004, the group controlled more than 1,300 radio stations – far exceeding the number of licenses held by any other company.[7] This trend towards deregulation and ownership liberalization is likely to continue, although public and Congressional pressure may serve to maintain an upper limit on some ownership combinations.

Issues surrounding industry structures and ownership are enduring and will continue to animate public policy discussions for the foreseeable future. As evidenced by the unprecedented critical public response to the FCC's proposed new ownership caps (described in Chapter 4), this is much more than an intramural debate between government regulators and industry giants. And because the media inhabit such an influential cultural universe, the stakes are high. In the twenty-first century the battle between those who favor a marketplace approach and those who favor some government regulation to prevent media content monopoly will be fought on both economic and ideological grounds. The battle could be intense.

Some critics complain that the issue is really one of a concentration of power and that such monopolistic tendencies stifle competition, which in turn results in decreasing content diversity. They argue that antitrust laws, which are intended to prevent unhealthy concentrations of power, have been ineffective sanctions against media consolidations. Still others see electronic media giants as a danger to democracy itself because of their perceived threat to both the marketplace of ideas and the diversity of views that sustain democratic pluralism. Media executives and other industry supporters respond that market forces should determine the configuration of ownership patterns and consumers' demand for choice will automatically motivate broad content variety. They also argue that, even with media consolidations, the United States exhibits the most diverse and competitive media system in the world.

New media can provide healthy competition for old, established media, forcing the latter to be wary of complacency. But over time, the new and the old tend to converge. Take, for example, the bitter rivalry that characterized the broadcast and cable industries from the inception of CATV. As we saw in Chapter 4, the FCC initially erected a protective economic shield around the established broadcast industry until it grudgingly repealed the most oppressive cable regulations. The broadcast industry, while still fearing cable television's competitive potential, gradually moved into alliances with cable,[8] although FCC regulations prohibited cable/broadcast cross ownership arrangements in *local markets*. However, a federal appeals court scuttled those rules in 2002, providing growth opportunities for companies such as AOL Time Warner and Comcast.

The two trends that will deserve government scrutiny in the Twenty-First Century are *cross-ownership* and *conglomeration*.[9] Ownership of the old and new media is now often in the same hands. Traditional broadcast companies, for example, own cable channels. Similarly, print media giants such as Time, Inc., have merged with electronic media powerhouses and crossed into the cable market. The big players in home video distribution have been such familiar names as ABC, CBS/Fox, and RCA/Columbia.[10] Most electronic media companies are part of large, often global corporate enterprises responsible to stockholders whose main interest is the bottom line,[11] a reality that will probably deter any substantive governmental attempts at reversing the concentrations of ownership within this "converged" media environment.

Conversely, supporters of deregulation are less concerned about the alleged dangers of unrestricted cross-media ownership than they are about governmental intrusion into the marketplace. They argue that content diversity will not suffer under such a laissez-faire regime because the marketplace is the ultimate determiner of the commercial success of media content, regardless of the source. Some media scholars, on the other hand, see dangers in unregulated cross-ownership patterns:

> Media cross-ownership on a mammoth scale raises democratic hackles. Unless the FCC and Congress carefully update and enforce cross-ownership rules to prohibit the existing media giants from combining print or broadcast interests with new media outlets in the same market, there is a potential for vast private monopolies of speech. Nothing in our free speech tradition prohibits government from legislating against such monopolies or semi-monopolies. Indeed, much of communications law – from common carrier regulations to the public trustee status of broadcasters – is a response to the need to control the private power of media giants such as Western Union or AT&T or RCA.[12]

Apart from media cross-ownership, critics also complain that the sheer size of media companies is disturbing in democratic nations that historically have celebrated dissent, unorthodox political views, and programmatic diversity. Conglomeration poses the question of the extent to which government regulators should intervene in the marketplace to protect energetic public discourse and robust competition in the conveying of entertainment and information.

The State of Electronic "Free Speech"

Government regulators – Congress and the FCC – are never far from center stage when certain kinds of objectionable content are transmitted through media that are subject to their control. But how can content regulation function in the age of convergence? The First Amendment, of course, severely limits government censorship, but the regulatory framework under which the FCC can review the performance of broadcast licensees has not been significantly altered, even in this era of deregulation. The Supreme Court's decisions affirming the FCC's authority to regulate broadcasting in the public interest remain sound Constitutional precedent – for broadcasting. As noted in Chapter 4, the rationale underlying this governmental authority resides in the physical limitations of the electromagnetic spectrum. But some commentators have observed that this rationale may have outlived its usefulness in light of advances in both technology and spectrum management.

The FCC is unlikely to return to the pre-1980s era of intense industry scrutiny, but offensive programming, such as indecency, will continue to engage the Commission's attention. Children's programming and the effects of televised violence on children will also continue to be a preoccupation of government regulators.. In the fall of 2004, for example, the FCC released new rules pertaining to children's programming, which included counting show promos as commercial time and prohibiting shows from linking to products pitched by their characters. Not surprisingly, various television networks and cable channels requested the FCC to delay implementation of these rules.[13]

The Internet has become the most recent focal point for concerns about media content. The Web, of course, is available to anyone with access to a computer, appears to be unmanageable from a legal perspective, and defies traditional national boundaries in its diffusion of content. One threshold issue is whether the Internet should be treated like the print media with full First Amendment protection or the broadcast media with its more narrow Constitutional protection. In today's limited case law, the Supreme Court has displayed a willingness to accord the Internet the full panoply of freedoms available to the print media.

Some legal experts are less worried about government censorship in Cyberspace than with the many challenges that the Internet poses to such traditional areas of law as libel, privacy, and the protection of *intellectual property* (such as copyright and trademark). Congress has immunized Internet Service Providers (ISPs) from liability in most cases because they are innocent conduits for messages transmitted through their sites. Despite the difficulties of policing the Internet, there is no reason to believe that traditional legal principles should not be applied to the vast realm of Cyberspace. The streaming of radio programming, for example, does not relieve stations of the legal obligation to pay licensing fees for the music they play. Similarly, the downloading of copyrighted music from file sharing services is illegal, as evidenced by the legal problems and subsequent demise of Napster.

Congress has taken a particular interest in *indecent content* on the Internet. With the Communications Decency Act of 1996, it criminalized the "knowing transmission" of any obscene or indecent material to recipients under the age of 18. The law stands as a classic example of government attempting to apply traditional and perhaps archaic regulatory schemes to a new medium. The language of the Communications Decency Act tracks very closely with the definition of indecency applied to broadcasting when in fact, the characteristics of the Internet and the broadcast media are quite distinct. Not surprisingly, the Supreme Court therefore held the law to be unconstitutional because of its vagueness, the absence of any significant rationale for applying broadcast standards to such new technologies, and because the statute was virtually unenforceable without prohibiting the dissemination of constitutionally protected material intended for adults.[14] The Court subsequently struck down part of a federal law intended to prevent the dissemination of "virtual" child pornography over the Internet[15] because virtual child pornography does not involve the use of real children, which is one of the rationales underlying the constitutionality of laws prohibiting child pornography in the first place.

As noted in Chapter 4, the Supreme Court's free speech principles have historically been predicated upon the proposition that different media may be treated differently under the First Amendment. Until 1987, for example, broadcasters were required under the fairness doctrine to provide access for opposing views on issues of public importance; a form of governmental editorial intrusion that would not be constitutionally tolerated for the print media. Increasingly, however, the constitutional distinction between the print and broadcast media has narrowed with the abolition of such policies as the fairness doctrine. In the twenty-first century, the convergence of old and new media will seriously erode any justification for anything other than a unitary First Amendment standard.

Two basic rationales can be advanced in favor of a free speech model that is based upon a marketplace functioning with a minimum of government interference. First, history demonstrates that government policies designed to foster "good journalism," "access by competing groups," or "program quality" tend to chill speech, not protect it. Such requirements may add more speakers to the mix, but they don't necessarily ensure greater diversity or quality. Second, any form of content-based requirement impermissibly involves government in the editorial decision-making process. In a sense, the government becomes a cultural gatekeeper in deciding what it believes to be in the best interest of the public.[16]

Critics of *deregulation*, of course, are not likely to silently retreat, and there will still be voices that campaign for expanded government oversight. They cite the competing principle to the unregulated marketplace model – the public interest standard – exemplified in the FCC's historical regulation of the broadcast industry. In this view, if broadcasters (or their new media successors) are left to their own devices, they will pander to the lowest common denominator, "decreasing the quality of important information while simultaneously increasing commercialization." Such programming, according to advocates of this position, "is not in the public interest and neglects the need to create an intelligent, civically active community where all citizens have access to the full range of information that they need for self-government."[17] Proponents of deregulation, of course, respond that the public interest model is paternalistic and, while it may produce more programming of a certain kind, it does not result in higher quality programming. A requirement of more programming targeted to children, for example, might fulfill some arbitrary quantitative standard, but the content may still fall close to the "lowest common denominator" in qualitative terms.

There is no reason to believe that the trend towards deregulation that began two decades ago will not continue to influence the formation of communications policy. Convergence will force the courts to rethink their traditional approach of dealing with each medium on its own terms for the purposes of First Amendment/free speech analysis. During the last century it was common to accept the duality of the *print* (no content regulation) and *broadcast* (regulation in the public interest) models with still a third regulatory model applied to the cable industry. The proliferation of new media and the convergence of old and new media have rendered these legal distinctions anachronistic. Nevertheless, government regulators, such as Congress and the FCC, do not intend to abandon the terrain entirely. For the foreseeable future there will continue to be some oversight of the traditional media. The nature of the Internet makes it impervious to most attempts at government supervision, but despite constitutional concerns, Congress will continue to monitor what it considers the most egregious aspects of cyberspace, namely pornographic and indecent Web sites to which children may have access.

Chapter Rewind

Our track record for predicting the social impact of new media and the form that the regulation of such media should take has not been stellar. History teaches us that old media are more likely to adapt to changed circumstances than just succumb to the advent of new media. Thus, traditional electronic vehicles, such as radio and television, will continue to function within the new media environment, but digital technology has made possible a convergence of old and new media that will eventually reshape the contours of regulatory policy. Old media are gradually (and in some cases not so gradually) morphing into the new media environment thereby blurring the lines of demarcation. This convergence will eventually render the old adage "different regulatory schemes for different media" anachronistic. Even the "spectrum scarcity" argument for regulation of the traditional broadcast media is less convincing now because of the plethora of competing electronic voices and because advances in technology have permitted more efficient use of, and alternatives to, that spectrum.

Nevertheless, the same concerns that have always preoccupied government regulators will endure in the twenty-first-century media environment. Issues involving concentration and conglomeration of ownership, oversight of certain kinds of offensive content, and First Amendment (free speech) philosophy as it pertains to regulation of both the old and new media will continue to figure prominently in electronic media regulation. The application of traditional rules to the new media, such as inflicting broadcast indecency standards on the Internet, is likely to be ineffective. But it is an open question as to what regulatory framework will eventually emerge in the new media environment.

SELF-INTERROGATION

1   Considering the regulatory environment in which the cable industry first came of age, why do you think cable failed to live up to its potential in the first two decades of its existence?

2   What is meant by media "convergence" and what regulatory challenges are likely to flow from this process?
3   Describe some of the most prominent emerging media systems of the twenty-first century. In what ways do these new media undermine the rationales for regulating the traditional media of radio and television in the public interest?
4   Describe three areas of enduring concern that are likely to occupy government regulators in the near future.

NOTES

1   Jeffrey Abramson, F. Christopher Arterton, and Gary R. Orren, *The Electronic Commonwealth: The Impact of New Media Technologies On Democratic Politics*. (New York: Basic Books, Inc., 1988), 7, 68.
2   Ibid., 7.
3   Ibid., 6.
4   See Roger L. Sadler, *Electronic Media Law* (Thousand Oaks, CA: Sage Publications, Inc, 2005), 97.
5   Yaron Katz, *Media Policy for the 21st Century in the United States and Western Europe* (Cresskill, NJ: Hampton Press, Inc., 2005), 132.
6   John Pavlik and Shawn McIntosh, "Convergence and Its Consequences," in Erik P. Bucy (ed.), *Living in the Information Age* (Belmont, CA: Wadsworth, 2005), 68.
7   Richard Campbell, Christopher R. Martin, and Bettina Fabos, *Media & Culture: An Introduction to Mass Communication*, 5th edn (Boston: Bedford/St. Martin's Press, 2006), 139–40.
8   See Abramson et al., *The Electronic Commonwealth*, 281.
9   Ibid., 280.
10   Ibid., 281.
11   Ibid., 72.
12   Ibid., 282.
13   "Networks Ask FCC to Delay Kids' Rules," *Broadcasting & Cable*, October 3, 2005, 14.
14   Reno v. American Civil Liberties Union, 520 U.S. 1113 (1997).
15   Ashcroft, Attorney General v. Free Speech Coalition, 535 U.S. 234 (2002).
16   See Stuart Minor Benjamin, Douglas Gary Lichtman, and Howard A. Shelanski, *Telecommunications Law and Policy* (Durham, NC: Carolina Academic Press, 2001), 971.
17   Ibid.

CHAPTER 9

Business Challenges

W. Lawrence Patrick

The US electronic media have grown from a few hundred radio stations in the 1930s to approximately 13,000 radio and television outlets today and from three television networks to over 250 broadcast and cable programming channels. With Web-streamed material, satellite radio and direct-to-video movies stirred into the mix, the electronic media are now exhibiting an insatiable appetite for content. Scripts cannot be written rapidly enough and productions are being consumed as fast as they can be shot and edited. Live and pseudo-live event programming has expanded to include minor sports, reality shows of every genre, talking head commentary programs and news programming beamed instantly from every corner of the globe.

Feeding the Media Machine

Always on the hunt for the next big hit and looking for creativity from sources other than traditional producers, television networks have turned to foreign networks and production houses for program ideas. American shows such as *Survivor, Who Wants to Be a Millionaire, Big Brother*, and *American Idol* are simply copies of programs that had already been successful in Europe. Conversely, American programs continue to be exported to other countries either as dubbed copies of the originals or as new productions based on the original American concept.

Wheel of Fortune, a simple game show created for the US first-run syndicated television market, has been broadcast as a locally re-made program in over 50 countries. For most American shows, such *international syndication* constitutes a significant revenue stream.

Even old television series and live action shows find new life on TV Land, The Game Show Network, ESPN Classic and Nick at Night. On these channels, old series and historic sports contests are replayed for those consumers who prefer such programs over today's offerings. Vintage movie cable networks such as AMC, Turner Movie Classics and The Western Channel satisfy consumers' desires for past cinema favorites.

New channels on the audio side have also opened up opportunities for many producers and directors who formerly had few outlets for their programs. The satellite radio networks, Sirius and XM, are permitting narrow music genres to be heard for the first time on a national basis. Mississippi Delta Blues, alternative artists and talk programming that could not find an audience on traditional media can now connect with new audiences.

Yet, business models are in conflict. While traditional record companies have fought for years to stop piracy and to collect payments for the copying of their intellectual property, new musical artists, hungry to establish themselves with fans, often provide free downloads of their music. They actually encourage consumers to download and share their songs in the hope of developing a core audience that will support them.

In 2005, for the first time, downloads of music surpassed CD sales. A whole generation is learning that music comes from the Internet and their friends rather than from established record stores. Brick and mortar sales outlets will steadily fade as music becomes a downloadable commodity. Record companies are looking at *encryption software* and *per use licensing* as methods to control the distribution of their intellectual property. These companies as well as radio stations are learning the hard truth that it is consumers who will determine what music will be disseminated and how and when they will use it.

Meanwhile, the advent of digital cameras, laptop computers with non-linear editing capability as well as the constantly expanding program marketplace are proving to be a media bonanza for independent production companies. Many cable networks, each with 168 hours per week to fill with programming, have created a production feeding frenzy. Independent producers are lining up at The History

Channel, Food Network, Discovery and other cable programmers with story ideas, series outlines and production budget proposals. For these producers, each new media channel offers new opportunities for placement of their shows.

Local cable channels now broadcast minor league, small college and high school sporting events. Networks such as Outdoor Living Network (OLN), have brought minor sports and outdoor living programs to television for the first time. Archival footage of wars, historic video of major construction projects, animal and science videos, travel and exploration footage and do-it-yourself home remodeling shows are all being turned into cable network series.[1] Reality shows spotlighting risk-taking, pungent human dramas, and bizarre situations now are broadcast nightly.

The media machine is constantly hungry for new approaches, high concept shows and the next great idea that will catch the attention of consumers and advertisers. Taste and behavioral boundaries are often pushed hard by some of these new programs. MTV's *Jackass*, with its stunts aimed at young males, certainly grabbed that demographic. This series proved so popular that it even spawned a motion picture. But despite warnings from the show's producers, it prompted many viewers to try dangerous stunts at home and this turned off sponsors.

When not too extreme, however, new reality concepts appeal to advertisers in several key ways. *Product placement* is now an important aspect of how companies try to reach consumers with the message about their wares. Programs such as *Survivor* regularly integrate automobile give-aways as part of the show's plot. Coca-Cola products were prominently displayed on *American Idol*. *The Apprentice* built each week's episode around specific companies and products. HGTV, with its home remodeling programs and annual home give-away, depends heavily on product placement revenues. Today, many advertisers are intimately involved in the development of new programs for both broadcast and cable television networks; going beyond product placement and conventional show sponsorship to creation of the program concept itself.

But as programming choices become more personal and video blogs or Internet video sharing become more popular, consumers are searching for programming alternatives that may not be advertiser-supported. The increasing consumer preference to interact with the media is pushing the programming envelope and creating a demand for new content that may sometimes not be satisfied through traditional business models.

Partners and Joint Ventures

As shows become more expensive and the media landscape more unstable, many media companies have learned that it is prudent to take on a *partner* before venturing into new productions and businesses. Since its beginning, the motion picture industry has been built on partnerships. Studios often loaned their stars to other studios for particular movies in exchange for a cut of the profits. This evidences a long-standing culture of "cooperative competition" among the motion picture studios. With executives, stars, writers and directors moving around Hollywood regularly, it is often beneficial to cooperate with today's competitor who may be tomorrow's ally.

In some cases, motion picture and television productions are simply too expensive for any one entertainment company to underwrite. With Hollywood blockbusters approaching $200 million to produce and with at least another $100 million more required for promotion and distribution, many studios believe that it is safer to share the upside incomes in order to also share the downside risks if the movie does not perform well. Universal may partner with Paramount or Columbia to produce and distribute a movie. Likewise, American motion picture studios often pre-sell the foreign distribution rights for a movie in order to help finance its production.

Another reason for partnerships and alliances between media companies is that different parties may bring different assets to a venture. For instance, studios have long relied on private investors and now seek out hedge funds (private capital investing institutions) to provide significant funding and financial expertise in order to bankroll new movies.[2] Content for a production also may come from a non-distribution studio, such as Pixar, that needs the distribution strength of its partner, Disney. At some point, such entities then may merge – exactly what happened in the case of Pixar and Disney.

Video games also have become the content for motion pictures. Suddenly, a gaming company is in an alliance with a motion picture studio to develop and promote the movie. For example, the director of *King Kong*, Peter Jackson, supervised the development of the *King Kong* video game that was released in conjunction with the movie. This game, which took more than a year to develop, cost over $25 million to produce. Such high production costs and the need to protect themselves from financial disaster are what leads to partnerships between media companies.

NBC launched its cable news channel MSNBC with partner Microsoft. HBO's *Rome*, an expensive and high-quality mini-series production, was co-produced by the British Broadcasting Corporation. That series debuted on HBO in America and was subsequently broadcast by the BBC in Britain. Time-Warner sold off 5 percent of its America On-Line subsidiary to Google, the successful search engine, for $1 billion. Google thus promotes AOL service while being featured prominently on the AOL site. Networks have regularly traded unsold on-air advertising for a piece of websites and Internet portals.[3] In short, partnerships are a common occurrence and a reality of life in today's media world.

Broadcasters and cable programmers also form partnerships to bid on sports rights packages and to underwrite new ventures before they are launched. The broadcast networks regularly consult their affiliates, and generally ask them to contribute cash and advertising inventory to support their bids for particular programming. With the cost of NFL rights, NASCAR and NCAA championships running into the hundreds of millions of dollars (or more) per year, the networks simply cannot shoulder the entire financial burden of these programming commitments. The affiliates want these programs because they often deliver some of the highest viewership levels of the season. Therefore, these stations have accepted the fact that they must help pay for this ultra-expensive product. This partnership helps both the networks and the affiliates while making the sports rights holders and team owners wealthy.

Some local television stations now also maintain local cable news channels or websites, often in conjunction with their local area's newspaper or cable system. Station employees therefore must know not only how to shoot, edit and present a news report on television but also how to write a newspaper story or create a more in-depth report that can be presented as part of a streaming video sidebar on the station's website.

Obviously, not every venture is made for a partnership. The egos of media executives, the inevitable creative differences over the direction of a movie or television production, and the fights over the equitable distribution of profits from a partnership often cause significant problems among media companies. In some cases, these fractures become permanent rifts between former partners. In other cases, such as the recent Viacom and CBS Corp. split, the end of the partnership was a strategic move designed to allow each company to grow faster and with less restrictions on it than when it was linked to the other.

Consumers and Advertisers

Broadcast media traditionally pushed content to consumers who enjoyed free consumption underwritten by advertisers. As they were rolled out, basic cable networks also relied on advertising without consumer objection. After all, consumers had been conditioned to advertising messages by watching decades of commercial broadcast television. Thus, cable networks such as ESPN, CNN, and MTV and others were able to tap a second revenue stream – advertising – to supplement the first stream of monthly per-subscriber payments from cable systems. Later, when Time Warner launched HBO, cable subscribers were introduced to the alternative concept of paying for programming that was delivered commercial-free. HBO was quickly followed by Cinemax, Showtime, Starz and other pay services. These channels lured consumers away from both broadcast television and basic cable programs that were advertiser-supported. And advertisers knew that every hour that consumers spent watching a pay cable movie was an hour where commercial messages could not reach these viewers.

The advent of *TiVo*, a hard-drive device capable of recording video programming and allowing consumers to fast-forward through commercials, struck another note of fear into advertisers' hearts. Would consumers skip advertising during playback? Some consumers report that they do. Other research, however, suggests the opposite. If consumers do record programs and then skip the commercials during playback, advertisers know that their messages are lost. These sponsors then resist the efforts of broadcasters to raise rates in light of evidence that their commercials are not reaching the intended audiences. The long-term impact of fewer viewers for commercials clearly concerns both advertisers and broadcasters.

Paradoxically, digital video recorders (DVRs) like TiVo also present advertisers with an opportunity by allowing them to produce long-form *infomercials* customized to the interests of individual consumers. These can be accessed by viewers via the Internet or stored on their DVR hard drives. Automobile manufacturers, already the top advertising category for television, now produce consumer-friendly half-hour infomercials about their new models. Viewers can download these messages for watching at their leisure.

New technologies also empower the consumer to move beyond the necessity of fast-forwarding through commercials and instead, pay for the programming up front. The television networks are now selling replays of some of their shows, absent commercial interruptions, for between 99 cents and $1.99 with the programs being available the day after their national broadcast.[4] The networks claim that this extends their reach to viewers who otherwise might not have viewed the program. And it certainly presents these networks with a new revenue opportunity even at the expense of advertiser penetration.

Conversely, the traditional media are exploiting the Internet as a way of building rather than dismantling the consumer/advertiser connection. Broadcasters accomplish this by using their reach to drive consumers to station websites. Once there, consumers are exposed to advertising tiles, click-through links to specific advertisers' own websites, and streamed advertiser-supported video. These streams can be replays of earlier broadcasts or new programming especially created for the web viewer.

Broadcasters have also learned to integrate product placement into programs. As discussed previously, product placement is now an important method for advertisers to promote the value of their product or service within the fabric of the program. Whether it is naming rights for a particular broadcast, a product give-away during the show or simply the inclusion and use of a product within a program, broadcasters and advertisers understand the value of product placement.[5]

As the mass media fragment and personal media become the norm, advertisers are re-inventing how they reach consumers. A smaller percentage of total advertising budgets will be spent on traditional media, while more dollars are diverted to more targeted technologies. *Data mining* and personalized electronic approaches are increasingly critical to marketplace success. Data mining is the process used by researchers and advertisers to determine what websites a person visits, what media the consumer uses, what purchases she/he makes and to uncover other demographic and economic data to define and classify prospective customers. Searching databases for prior purchase information and assumptions about spending habits based on a person's home location provides advertisers with clues on how best to target particular consumers. The goal is to determine the interests and orientations of each consumer and then tailor messages designed to reach them.

For instance, if an advertiser can ascertain that certain consumers enjoy home improvement programs such as those on HGTV, that they spend time on websites dealing with home improvement issues, have made prior purchases at a Home Depot or Loew's and live in a neighborhood where home remodeling is common, then the advertiser may be able to target advertising messages to these consumers more effectively than by using more generalized mass media approaches.

In order to make up for the decline of mass delivery systems, advertisers are also injecting their brands and messages into new electronic venues. From electronically-inserted messages on playing fields, to video displays in elevators and shopping carts to ads on water coolers and imbedded in store floors, advertising surrounds consumers in ever-widening ways. Electronic marketers must now be expert in far more media than radio and television. And the operators of media systems must be ready to assist these marketers in executing their cross-media communication strategies.

Restructuring the Media Landscape

As media companies strive to cope with the reality of personal media, they are learning valuable lessons about how to assemble consumers, deliver content to these consumers and generate revenues from these activities. Radio and television advertising sales will continue as a profitable business well into the foreseeable future. Although the audience levels may be lower, broadcasters will still be able to raise their advertising rates and generate substantial revenues. Profit margins for many television broadcasters may erode from the longstanding 40 and 50 percent levels to more modest 20 and 30 percent levels. Nevertheless, these returns are still well above those posted by most businesses. The re-purposing of content through Internet downloads of programs to home computers or iPods will provide some incremental revenues for traditional media companies and, to some extent, shore up declining profit margins that rely on advertising.

As discussed previously, operating a television network is not a high profit business. Running local television stations, on the other hand, is very profitable. Therefore, as the pressure on broadcast networks from cable, personal media and the Internet increases, the networks will continue their push to raise station ownership caps in order to protect their profitability by enlarging their O & O groups. In 2006, all four of the major television networks were already bumping up against the current 39 percent ceiling on national audience reach. This cap is the percentage of total national audience reach that one company can achieve through its ownership of television stations. The FCC's original 35 percent national cap was raised by Congress in 2004 to 39 percent in a compromise over the Commission's plan to raise the cap to 45 percent. The networks are in favor of an increasingly higher cap while the affiliate groups are opposed to any increase in the cap that would increase the networks' clout at their expense.

How large media companies can become is a fundamental issue in the restructuring of the media landscape. As illustrated in Table 5.1, the seven big media giants all operate on multiple platforms across many technologies. The four of these companies that own television networks certainly favor relaxed regulation that would allow them to control more television stations.

Another way to restructure the media landscape is to increase *cross-national ownerships*. Already, some American media companies control radio and television stations, movie theaters, cable networks, newspapers and other media overseas. As regulatory barriers are removed that prohibit such ownership, more American companies are likely to expand their investments in foreign station ownership. They do this for two reasons. First, there are not enough large stations to purchase in the United States. Even if they have significant dollars to spend, there are no stations in the top 25 markets to buy. Second, as consumer product companies become multi-national (think world-wide brands like Coca-Cola, McDonald's and Sony), American media companies want to be able to offer these advertisers multi-national advertising platforms to promote their products.

In recent years, broadcasters have also expanded into non-regulated domestic media such as outdoor advertising, in-store entertainment, concert promotions and event sponsorship. They do this because, though the profits generated by broadcasting operations are substantial, there is a limit as to how many additional stations are available for purchase. So in many cases, monies generated by broadcasting are best spent on growing related media businesses. Faced with enormous profits and the taxes that would be due on them, most broadcast interests buy whatever additional properties are available rather than paying taxes on profits sitting in the bank.

But perhaps the most significant change for broadcasters in the coming years will be the shift from analog to digital television and the multiple channels that one digital allocation will support.

It is unclear what broadcasters will place on these new channels. So far, most television stations are simply transmitting their basic analog programming on the first digital channel and then using the rest of their digital spectrum allocation for weather radar, traffic cameras, time-shifted repeats of current programming or a secondary network, such as the CW or a Spanish-language network. NBC Television was the first network to create a second video network feed for its affiliates specifically designed to be carried on a digital channel.[6] This channel was designed as a fulltime national weather service with local insertions. The question of what programs and services other broadcasters will use to fill these additional avenues is a critically important one. At the moment, the technology of creating additional channel capacity is running far ahead of the programming content to fill these new digital slots.

Local news channels and regional networks may be given *national* distribution as broadcasters search for more programming volume. For example, smaller cable networks such as TV One, a service designed for African-Americans, could find national carriage on broadcasters' extra digital channels as an alternative to seeking carriage on traditional cable systems. Cable operators often ask new programmers for a piece of ownership in exchange for assisting them with their launch. Discovery Networks, The History Channel and others have had to give up a part of their equity in order to secure national cable carriage agreements with the leading multiple system operators. But now, broadcast digital television, where every station can suddenly air six channels rather than one (and this may soon increase to 10 or more channels depending on bandwidth compression advances), becomes a new distribution option for emerging program services.

As discussed in Chapter 6, the government has pushed for this shift from the analog to digital platforms so that it may reclaim the analog spectrum that broadcasters have historically occupied. Faced with growing demand from cell phone and other mobile telecommunications providers as well as from emergency first-responder users, the

FCC needs to reallocate spectrum occupied by these old analog television channels. The auctioning off of this spectrum, once it is returned to the government by the broadcasters, is expected to bring between $7.5 and $10 billion to the Federal Treasury.[7] For its part, Congress sees this move to digital television not only as an important revenue generator but also as a means of providing emergency responders with access to communications spectrum. It hopes that expanded emergency communication systems can avoid the breakdowns experienced in the wake of the 9–11 attacks and Hurricane Katrina.

The media landscape, both for consumers as well as for broadcasters and cable operators, will be very different in coming years. Technology continues to provide personal media choices as well as dazzling new opportunities for entertainment and information. But the business models to support these new services are not always readily apparent to old-line media companies. They are often still focused on traditional means of production and distribution. It may be a temporary generational issue as senior executives in charge of those companies who are not well acquainted with new technologies and consumer behavioral changes are replaced by younger managers.

Those companies that are agile, work closely with domestic and international partners to broaden their opportunities, and embrace the new technologies as brand extensions of their current offerings will do well. Others, who ignore the opportunities and/or do not grasp how consumers are now using media, will fall by the wayside.

Living in a Personal Media World

As the mass media fragment and both information and entertainment become ubiquitous, consumers are faced with many choices. Information flow is quicker. There are more content options and consumers must develop filters to protect themselves from sensory overload. With 500 channels of cable or satellite television, consumers need help simply in knowing what is on at a particular time and what might be of interest to them. They also need filters to protect them from Internet spam, infomercials they do not want to view and product placement messages disguised as entertainment.

Consumers also find themselves spending much more time organizing their media usage and making selections than in previous years when they passively ingested what broadcasters decided to send them. Whether it is loading their iPods, using video on demand recorders, or employing computer search engines to find information, consumers engage with media in ever more varied ways. Admittedly, some don't like coping with new devices or choosing among a vast array of program options. Rather than embracing the new world of program and activity choices, they long for the days of linear programming dictated and delivered by the networks. But those days are slipping away.

Technology promises to make the pace of adoption of new devices even more rapid. Cell phones are a good example. In less than a decade, cell phones have gone from large, bulky objects possessed by only a few to inexpensive, lightweight devices owned by every suburban teenager who uses the phone to take pictures, instant message with friends and download Internet offerings. Consumers find that their choices of video and audio programming are multiplying at an exponential rate and this programming has become ever more portable. We have come a long way from the AM-only transistor radio Zenith marketed as a personal communications breakthrough in 1962 (see Figure 9.1).

These new choice and access options raise serious concerns about information overload, as well as content and privacy issues. Parents are worried that portable computers and handheld wireless devices that can surf the Web may give children access to "adult" information and entertainment. Despite efforts to block such content, technology presents gateways that even the most vigilant parent cannot barricade. Meanwhile, with the growth of video distribution and the ability of individuals to digitally shoot and edit their own programs, questions of privacy become critical. Personal video blogs, an Internet version of home movies, can transmit images and content far beyond one's immediate neighborhood. The damage from privacy violations can be severe.[8] The illegal sale of digital credit information is a common crime in today's world. So, unfortunately, are digital pictures taken without the subject's permission (the *"up-skirting" danger*), and the fact that such personal information can be posted on the web and instantly be transmitted worldwide. This immediate and massive danger to one's privacy is a notable downside to the new digital world.

Figure 9.1 Zenith's all-transistor Royal 500-E "pocket" AM receiver. The most powerful radio of its size when introduced in 1962. Source: Courtesy of Zenith Electronics Corporation.

All these personalized recording and distribution devices also raise questions about the protection of intellectual property. MP3 players and home computers have demonstrated that copyrighted music can be ripped and swapped among consumers with no thought given to the copyright violations involved. While both artists and media companies view technology as a method to expand the reach for their products, they also see it as a threat to their very existence. For new artists, Internet distribution of their music or video programming gets them exposure and the ability to create a fan base. But for established artists, new technology must be harnessed so as to allow the artist as well as the recording studio to capture revenues from the popularity they already possess.

What's Ahead

What's ahead is change. And change at an ever-increasing pace. The changes will be in how the media must present their products and programs to consumers. There will also be changes in how advertisers will reach consumers with their message. Perhaps most importantly, there will be vast changes in how the media are used by consumers. The pace will only accelerate in coming years and media companies will either move quickly to capture new opportunities or be left at the curb as the parade passes them by.

For media companies, the goal must be to expand their reach as well as to understand new business models. Networks are realizing that having a dominant web presence is critical.[9] The networks are also learning to integrate web or cell phone technology into their programs. This is perhaps best seen by the text-voting for winners in programs such as *American Idol* or *Nashville Star* and instant chances for viewers themselves to become winners on shows like *Deal or No Deal*. Being able to promote products in a program and then have consumers purchase those products through a show or network website is also an example of cross-platform promotion and revenue generation. Mobile television viewing on cell phones, outdoor displays and transit video display terminals is now part of the media landscape.

Traditional advertising models are faltering. Video on demand options with the associated ability to skip commercials threaten traditional media companies. Advertisers now target consumers via the Internet as well as non-traditional in-store video displays. Technology has allowed companies to generate targeted pitches to consumers far more effectively than in the past and at a substantially lower price. Marketers can send out email offers to select individuals rather than placing expensive network commercials that blanket customers and non-customers alike. All these developments resulting from the new world of digital media threaten the traditional media's economic base.

Broadcasters must learn that one of their most powerful promotional strategies may be to drive consumers to advertisers' websites where the real sales pitch can more expansively be made. Similarly, websites for local television stations are the new battleground for revenues and profits.[10] Traditional advertisers still use the mass media but now spend an increasing portion of their budgets on non-traditional media. Broadcasters can re-capture some of these monies both by hosting interactive websites for their stations and by serving as portals to the websites of their advertisers.

Tomorrow's electronic media executives must understand not just traditional broadcasting, but every new technology and how consumers are using it. This involves everything from downloading programs onto iPods, to exploiting cell phones and emails for sending news and program alerts, to building successful web advertising models. It also encompasses product placement opportunities, producing long-form video infomercials, and joint venturing with a large number of marketing and media partners. Electronic media professionals must be able to determine how best to target tomorrow's consumers and which methods can reach them effectively and efficiently. This may mean combining traditional television commercials with email approaches and web advertising tied to the consumer's interests and geographic location.

For their part, broadcasters have always made *localism* the bedrock of their success. Their ability to inform the community about what is happening in their neighborhoods has been a key component of this success. Localism is the element that broadcasters must continue to stress. Other media may be able to present targeted informational or advertising content to consumers, but local television and radio stations continue to be the best source of news, weather, traffic and information. For the new media operators, finding a way to establish strong community-focused content is the best way to challenge traditional delivery systems.

New challenges bring new opportunities. Some companies and executives will embrace each new development, study it and learn how to make it work for their businesses. Others will be slow to react, will make poor choices and will be left to struggle for survival. The business challenges for tomorrow are complex, require financial intelligence, and some level of risk-taking by media companies. Executives should regularly be ready to try, and then either expand or abandon, new business opportunities as they attempt to steer their media companies through the waters of change.

Chapter Rewind

The explosion of cable networks, broadband distribution and the Internet has caused a rapid increase in the amount and scope of video production. Niche networks, targeting very specific consumers, have shown that they have drawing power. Some of these programs are literally built around product placements with producers working closely with advertisers as each show is developed.

Having partners and joint ventures to produce programming is essential in today's hyper-competitive world. Few companies wish to assume the risk of launching new cable networks or large-scale productions on their own. Similarly, the costs of sports rights has increased to the point where game packages are split up among broadcast and cable networks.

Advertisers are struggling to figure out how best to target consumers and reach them through both traditional and new media. Technology such as DVRs has allowed consumers to skip commercials. Conversely, such devices permit advertisers to provide program-length infomercials that can entertain while still presenting the benefits of their products.

The rise of new media has re-defined how traditional broadcasters operate. Regulatory fights between networks and affiliates will continue. Local broadcasters are learning that appealing local news content as well as website marketing ability are critical to their survival. The complete conversion to digital television will usher in greatly expanded offerings by television stations.

Cell phones, video iPods, and television screens that look and act more like large computer monitors will change how people use media. Change is the constant in the new media world. Consumers will constantly re-define how they use the media, the programming that they want and the technology that they select to obtain it. Media executives will need to be agile and creative to stay ahead of this constantly advancing curve.

SELF-INTERROGATION

1   Explain the increasing demand for television programming as well as the concept of niche programming.
2   What is product placement and how has it changed the ways in which advertisers are involved in television programs?
3   In the era of new media, why do programming and distribution companies seek out partners and joint venture arrangements? What are the advantages and disadvantages of this approach?
4   How have the jobs of advertisers changed in the world of new media?
5   How do advertisers reach more mobile consumers who use media personally?

6   Provide some examples of how network television shows are re-purposed. What problems does this practice cause for the networks and their affiliates?

7   Describe the battle between consumers who want free access to content and the corporations who own intellectual property and are trying to protect it. How do aspiring and established artists figure in this battle?

8   What does the future look like for local television broadcasters?

9   Describe what station owners should be doing to protect their franchises and insure their continued profitability.

NOTES

1   Discovery, Animal Planet, The History Channel, The Travel Channel, HGTV and Fine Living are networks that feature this type of programming. They consume enormous amounts of original product annually.

2   For a good discussion of private financing for motion pictures, see Gabriel Snyder, "Other People's Money," *Variety*, January 23–29, 2006, p. 1.

3   Mel Karmazian, the former head of CBS and later top executive at Sirius Satellite Radio, regularly traded on-air advertisements on the CBS and UPN networks, billboard space from Viacom Outdoor and radio advertisements on the company's Infinity radio stations for financial stakes in nearly three dozen Internet portals. Some of these included the very successful WebMD and Marketwatch.

4   CBS has recently announced that it would begin downloading some of its entertainment programs for free and without commercials to increase the reach of the programs.

5   Product placement has also drawn the attention of the Federal Communications Commission. The FCC has warned broadcasters that such product placements must be disclosed. There are also specialized companies whose business revolves around obtaining product placement for their consumer goods clients. These firms work with networks, movie studios, production companies and the talent themselves to include references to their products in television, movies and music.

6   This service is called WeatherPlus. It took two years of negotiating between the network and its various affiliate groups before launching this service in 2004.

7   The FCC is one of the few federal agencies that produces significant revenues. It achieves this by auctioning off spectrum rights to various industries. The analog television spectrum is extremely valuable for cell phone companies who wish to invest in G4 (generation four) advanced cell phone services. These services, which would allow mobile Internet

platforms, video cameras in cell phones, and make the cell phone a portable computer, require an expanded spectrum in which to operate.

8   This is particularly true with inappropriate content being spread via the Internet. With a few keystrokes, video downloads and pictures can be available worldwide. Law enforcement officials are particularly concerned with the growth on the Internet of child pornography and privacy invasion from "up-skirting" producers. Even terrorist organizations are adept at using the Internet to distribute information about their causes and to contact worldwide sleeper cells via video and data technology.

9   News Corporation, for instance, has aggressively purchased web sites and Internet portals in an effort to reach consumers in new ways and to be able to cross-promote their programs and products. By 2006, the company's investment in these sites was in excess of $500 million.

10 Some broadcasters are realizing annual seven-figure revenues from their web sites by streaming videos for clients and providing links on those sites to the web sites of advertisers. Many stations now embrace this technology, exploiting it as a primary business extension of their advertising sales effort rather than a side-line. Such stations are regularly experiencing double-digit growth for their Internet ventures.

Index

Note: Terms are indexed by their acronym when that acronym is more commonly used than the complete phrase for which it stands.